ENCYCLOPAEDIA OF ANIMAL DISEASES - IV

PHYSIOLOGICAL DISEASES

By

Ashok Kumar

Department of Zoology
Bundelkhand University
Campus Department
Jhansi (India)

DISCOVERY PUBLISHING HOUSE PVT. LTD.
NEW DELHI-110 002

Published by:
Namit Wasan

DISCOVERY PUBLISHING HOUSE PVT. LTD.
4383/4B, Ansari Road, Darya Ganj
New Delhi-110 002 (India)
Phone : +91-11-23279245; 23253475; 43596065
E-mail : discoverybooksindia@gmail.com
discoverypublishinghouse@gmail.com
namitwasan9@gmail.com
web : www.discoverypublishinggroup.com

Edition: **2020**

ISBN: 978-81-8356-206-5 (Set)

ISBN: 978-81-8356-285-0

Physiological Diseases

Printed at:
Infinity Imaging Systems
Delhi

PREFACE

The **Physiological Diseases** has been carefully compiled and edited to meet the long felt needs of increasingly large number of those who have to deal with the different aspects of human diseases in colleges, universities and research institutes. It provides a stimulating and important new view of interaction between animals and pathogens causing diseases. The objective is to introduce to students the essential principles for understanding various aspects of diseases. Most of diseases constitute the largest part of human pathology and are the primary cause of death. Hence, special importance is given to the study of such diseases.

The book is intended to acquaint students of various fields involved directly or indirectly with the major principles of human diseases. The book may be helpful as well to practitioners and those engaged in medical research.

In the preparation of this book large number of books and research papers have been consulted. So no authenticity is claimed.

The author wishes to express his deepest appreciation to the many people who have contributed in one way or the other to the preparation of this title.

The author expresses his gratitude to Mr. Wasan and staff of M/s Discovery Publishing House for their whole hearted co-operation in the publication of this book.

The author tried hard to be accurate and upto date in statement and realises the impossibility of completely avoiding errors therefore, the author will greatly appreciate having his attention called to any questionable statement.

Author

CONTENTS

RESPIRATORY SYSTEM

MAIN CLINICAL SYNDROMES

Syndrome of Focal Consolidation of Pulmonary Tissue

The syndrome of focal cosolidation of lung tissue is caused by filling of the alveoli with the inflammatory fluid and fibrin (in pneumonia), blood (in lung infarction), growing connective tissue in the lung (pneumosclerosis, carnification) in long-standing pneumonia, or developing tumour. The common complaint of the patient is dyspnoea. Examination of the patient reveals thoracic lalgging of the affected side during respiratoin; vocal fremitus is intensified in the consolidated area; the percussion sound over the consolidation site is slightly or absolutely dull; auscultation reveals bronchial respiration, exaggerated bronchophony and (in the presence of liquid secretion in fine bronchi) resonant (consonating) rates. X-ray examination shows the focus of consolidation as an area of increased density in the lung tissue, its size and contours depending on the character and stage of the disease, and some other factors.

Cavity in the Lung

Cavity in the lung is formed in abscess or tuberculosis (caveren) or during degradation of the lung tumour. An empty large cavity is communicated with the bronchus and surrounded by a ring of inflamed tissue. Examination of the chest reveals unilateral thoracic lagging and intensified vocal fermitus. Percussion reveals dulled tympany or (if the cavity is large and peripheral) tympany with a metallic tinkling. Auscultation reveals amphoric breathing, intensified bronchophony, and often medium and large resonant vesicle rates. X-ray examination proves the presence of the cavity in the lung.

Fluid in the Pleural Cavity

The syndrome of accumulation of pleural fluid occurs in hydrothorax (accumulation of non-inflammatory effusion, i.e. transudate, for example in cardiac failure), or in pleurisy with effusion (inflammation of the pleura). The syndrome is characterized by dyspnoea due to respiratory insufficiency caused by lung compression and decreased respiratory surface, asymmetry of the chest (enlargement of the side where pleural effusion is accumulated) and unilateral thoracic lagging during respiration. Vocal fremitus is markedly weakened over the area of the pleural effusion or it may be undeterminable; percussion reveals a dulled are markedly weakened or absent. X-ray examination reveals an area of increased density in the area of accumulation of the pleural a fluid, which is usually at the bottom of the chest (often bilateral in hydrothorax). Its upper border is quite distinct. If transudate is accumulated in the pleural cavity its border is more horizontal, while in the presence of pleural effusion, the border is scant, to coincide with the Damoiseau's curve as determined by percussion.

Air Accumulation in the Pleural Cavity

Air is accumulated in the pleural cavity when the bronchi are communicated with the pleural cavity (in subpleural tuberculosis cavern or abscess), in injury to the chest, or in artificial pneumothorax (injection of air into the pleural cavity for medical purposes in the presence of large caverns in the lungs). Asymmetry of the chest found in this syndrome is due to the enlarged side where air is accumulated; the affected side of the chest cannot take part in the respiratory act. Vocal fremitus is markedly weaker or absent altogether over the site of air accumulation; percussion reveals tympany. Breathing sounds and bronchophony are either weak or absent and are not conducted to the chest surface to be detected by ausculation. X-ray examination reveals a light pulmonary field without pulmonary pattern; a shadow of the collapsed lung can be seen toward the root.

External Respiratory Dysfunction

The function of the external respiratory apparatus is to supply the body with oxygen and to remove carbon dioxide formed by exchange reactions. This function is realized firstly by ventilation, i.e. gas exchange between the outer and alveolar air. This ensures the required oxygen and carbon dioxide pressure in the alveoli (an important factor is intrapulmonary distribution of the inspired air). Secondly, this function is realized by diffusion of carbon dioxide and oxygen through the walls of the alveoli and lung capillaries (oxygen is supplied from the alveoli to the blood and carbon dioxide is diffused from the blood to the alveoli). Many acute and chronic disease of the bronchi and the lungs cause respiratory insufficiency. The degree of morphological changes in the lungs does not always correspond to the degree of their dysfunction.

Respiratory insufficiency is now defined as the condition with abnormal gas composition of the blood, or this abnormality is compensated for by intense work of the external respiratory apparatus and higher load on the heart. This decreases functional abilities of the body. It should be noted that the external respiratory function is closely connected with the blood circulatory function: the heart work is intensified during external respiratory insufficiency, which is an important compensatory element of the heart function.

Respiratory insufficiency is manifested clinically by dospnoea and cyanosis; at later stages, when cardiac failure joins the process, oedema occurs.

The patient with respiratory insufficiency employs the same compensatory reserves as a healthy person does during heavy exercise. But the compensatory mechanisms of a sick person are actuated much earlier and at loads under which a healthy person would feed no discomfort (e.g. dyspnoea and tachypnoea can develop in a patient with lung emphysema even during slow walking).

Among the first signs of respiratory insufficiency are inadequate changes in ventilation (rapid and deep breathing) at comparatively light loads for a healthy individual; the minute volume increases. In certain cases (bronchial asthma, lung emphysema, etc.) respiratory insufficiency is compensated by intensified work of the respiratory muscles, i.e. by the altered respiratory mechanics. In other words, in patients with pathology of the respiratory system, the external respiratory function is maintained at the required level by mobilizing compensatory mechanisms (i.e. by efforts greater than required for healthy persons), and by minimizing the respiratory reserves: the maximum lung ventilation decreases, the coefficient of oxygen consumption drops, etc.

Various mechanisms are involved gradually to compensate for progressive respiratory insufficiency depending on its degree. At the early stages of respiratory insufficiency the external respiratory function at rest is realized in normal way. The compensatory mechanisms are only actuated during exercise in a sick person. In other words, only reserves of the external respiratory apparatus are decreased at this stage. As insufficiency further progresses, tachypnoea, tachycardia, and signs of intensified work of the respiratory muscles (during both inspiration and expiration), with involvement of accessory muscles, develop during light exercise and even at rest. At the later stage of respiratory insufficiency, when the body compensatory reserves are exhausted, arterial hypoxaemia and hypercapnia develop. In addition to the growing vivid arterial hypoxaemia, signs of latent oxygen deficit also develop; underoxidized products (lactic acid, etc.) are accumulated in the blood and tissues.

Still at later stages, right ventricular incompetence joins pulmonary insufficiency because of the developing hypertension in the lesser circulation, which is attended by increased load on the right ventricle, and also because of dystrophic changes in the myocardium occurring as a result of its constant overload and insufficient oxygen supply. Hypertension in the vessels of the lesser circulation in diffuse affections of the lungs arises by reflex mechanisms in response to insufficient lung ventilation and alveolar hypoxia — the Euler-Liliestrnd reflex (this rflex mechanism is an important adaptation means in focal lung affections; it limits blood supply to insufficiently ventilated alveoli). Further, in chronic inflammatory diseases of the lungs due to cicatricial and sclerotic changes in the lungs (and due to affections in the lung vessels) blood passage through the lesser circulation becomes even more difficult. Increased load on the myocardium of the right ventricle stimulates gradual development of its insufficiency to cause congestion in the greater circulation (pulmonary heart).

Depending on the cause and mechanism of developing respiratory insufficiency, three types of disordered lung ventilation are distinguished: obstructive, restrictive and mixed (combined).

The obstructive type is characterized by difficult passage of air through the bronchi (because of bronchitis, bronchospasm, contraction or compression of the trachea or large bronchi, e.g. by a tumour, etc.). Spirography shows marked decrease in the MLV and PVC, the VC being decreaed insiginficantly. Obstruction of the air passage increases the load on the respiratory muscles. The ability of the respiratory apparatus to perform additional functional load decreases (fast inspiration, and especially expiration, an also rapid breathing become impossible).

The restrictive type of ventilation disorder occurs in limited ability of the lungs to expand and to collapse, i.e. in pneumosclerosis, hydro- and pneumothorax, massive pleural adhesions, kyphoscoliosis, ossification of the costal cartilages, limited mobility of the ribs, etc. These conditions are in the first instance attended by a limited depth of the maximum possible inspiration. In other words, the vital capacity of the lungs decreases (together with the maximum lung ventilation), but the dynamics of the respiratory act is not affected: no obstacles to the rate of normal breathing (and whenever necessary, to significant acceleration of respiration) are imposed.

The mixed, or combined type includes the signs of the two previous disorders, often with prevalence of one of them; this type of disorder occurs in long-standing diseases of the lungs and the heart.

External respiratory dysfunction occurs also when the anatomoical dead space increases (in the presence of large cavities inside the lungs, caverns, abscesses, and also in multiple large bronchiectases). Similar to this type is the respiratory insufficiency due to circulatory disorders (e.g. ini thromboembolism, etc.) during which part of the lung is excluded from gas exchange, while its ventilation is to a certain degree maintained. Finally, respiratory insufficiency arises during uneven distributionof air in the lungs (distribution disorders), when a part of the lung is not ventilated (in pneumonia, atelectasis), with preservation of blood circulation. Part of venous blood is not oxygenated before it enters pulmonary veins and the left chambers of the heart. Similar to this type of respiratory insufficiency (with regard to pathogenesis) is the so-called vascular bypass or shunting (from regard to pathogenesis) is the so-called vascular bypass or shunting (from right to left), during which part of the venous blood from the pulmonary artery system enters directly the pulmonary vein (bypassing the capillaries) to mix with oxygenated arterial blood. Oxygenation of blood in the lungs is thus upset but hypercapnia may be absent due to compensatory intensification of ventilation in the intact parts of the lung. This is partial respiratory insufficiency (as distinct from total insufficiency where hypoxaemia and hypercapnia are present).

Respiratory insufficieny is characterized by upset gas exchange through the alveolar-capillary membrane of the lungs. It occurs when this membrane is thickened to interfere with normal gas diffusion through it (the so-calaled pneumonoses, alveolar-capillary block). It is not accompanied by hypercapnia eithe rsince the rate of CO_2 diffusion is 20 times higher than that of oxygen. This form of respiratory insufficiency is, in the first instance, characterized by areterial hypoxaemia and cyanosis. Lung ventilation is intensified.

Respiratory insufficiency associated with toxic inhibitionof the respiratory centre, anaemia, or oxygen deficit in the inhaled air, is not connected directly with the pathology of the lungs.

Acute and chronic respiratory insufficiency are differentiated. The former occurs in attacks of bronchial asthma.

Three degrees and three stages of respiratory insufficiency are also distinguished. The degrees of respiratory insufficiency reflect the gravity of the disease at a given moment. The first degree of respiratory insufficiency (dyspnoea, in the first instance) becomes evident only at moderate or significant physical load. Dyspnoea develops during light exercise in the second degree of insufficiency; the compensatory mechanisms are involved when the patient is at rest and functional diagnosis can reveal some deviations from the normal indices. The third degree is characterized by dyspnoea at rest and cyanosis as a manifestation of arterial hypoxaemia; deviations from the normal indices during functional pulmonary tests are significant.

Stages of respiratory insufficiency in chronic diseases of the lungs reflect the changes occurring during the progress of disease. Stages of latent pulmonary, pronounced pulmonary, and cardiopulmonary insufficiency are normally differentiated.

Treatment

The treatment of respiratory insufficiency includes (a) treatment of the main disease upon which the insufficiency depends (pneumonia, pleurisy with effusion, chronic inflammations in the bronchi and the pulmonary tissue, etc.); (2) removal of bronchospasm and improvement of lung ventilation (giving broncholytics, remedial exercises, etc.); (3) oxygen therapy; (4) pulmonary heart is treated by cardiac glycosides and diuretics; (5) phlebotomy is indicated in the presence of reflex erythrocytosis and congestion in the greater circulation.

SPECIAL PATHOLOGY

Among the vey great number of various diseases of the respiratory organs, most common are inflammatory affections of the bronchi (acute and chronic bronchitis, bronchiectasis), of pulmonary tissue (acute and chronic pneumonia, less frequently destructive affections such as abscess of gangrene of the lungs), and of the pleura (pleurisy, empyema). The lungs are affected selectively only in tuberculosis. The incidence of tuberculosis was high in old times. Certain occupational hazards (silicons and the like) can cause special forms of inflammatory diseaes of the lungs provided safety measures are not taken and the exposure becomes chronic. Disease with a protracted course, and also inadequately treated inflammatory disease (or in decreased reactivity of the body) can terminate in focal or diffuse pneumosclerosis. Frequent are tumours of the lungs (bronchogenic cancer, metastases into the lungs of cancer tumours of other organs). Pulmonary vessels are also often affected (embolism, lung infarction); these affections may also include affections due to foreign bodies of the bronchi, injuries to the lung (contusion, compression, etc.), congenital abnormalities (cysts, etc.), and genetically determined systemic diseases with lung affections (mucoviscidosis, deficit of alpha-1-antitrypsin, etc.); comparatively often occur allergic diseases (bronchial asthma), and many other pulmonary pathologies.

BRONCHITIS

Bronchitis is inflammation of the bronchi. The disease stands first in the list of respiratory pathologies and occurs mostly in children and the aged.

Primary and secondary bronchitis are distinguished. In primary bronchitis, inflammation develops as the primary process in the bronchi. Secondary bronchitis attends other diseases, such as influenza, whooping cough, measles, tuerculosis, chronic diseases of the lungs and the heart. According to the type of the inflammatory reaction bronchitis is subdivided into catarrhal, mucopurulent, purulent, fibrinous, and haemorrhagic. Bronchitis can also be focal or diffuse, depending on the localization of inflammation. Inflammatory process can reside only in the trachea and large bronchi (tracheobronchitis), the bronchi of the fine and medium calibre (bronchitis), or in the bronchioles (bronchiolitis, occurring mostly in infants). According to the course of the disease, acute and chronic bronchitis are differentiated.

Acute Bronchitis

Aetiology and pathogenesis

Acute bronchtis can arise (1) in acute infectious diseases, such as influenza, whooping cough, measles, and the like; (2) in chills which decrease the body's resistance to microbes (saprophytes) that are present in the upper air ways, i.e. pneumococci, pneumobacilli (Klebsiellae), streptococci, etc. Mechanical and chemical factors, such as coal, cement or lime dusts, formaldehyde, acids, or acetone may promote the onset of the disease. Asthenia (especially after grave diseases) and chronic sinusitis and rhinitis are also predisposing factors to the development of acute bronchitis.

Pathological anatomy

The disease begins with hyperaemia and swelling of the ronchial mucosa, hypersecretion of mucus and diapedesis of leucocyes; then follows desquamation of epithelium and formation of erosions; in grave bronchitis, inflammation may invovle the submucous and muscular layers of the bronchial walls and peribronchial interstitial tissues.

Clinical picture

The patient feels discomfort in the throat and retrosternal smarting. The voice becomes hoarse, the patient feels weakness, and excess perspiration develops. Cough is first dry or with expectoration of scant tenacious sputum; cough may be coarse, resonant, sometimes barking; it may come in excruciating attacks. Sputum is expectorated on the second or the third day of the disease. First it is mucopurulent, sometimes with streaks of scarlet blood; then it becomes purulent; coughing gradually subsides and softens.

The body temperature in focal bronchitis is normal or subfebrile; in grave diffuse bronchitis it may rise to 38-39°C. The respiration rate increases insignificantly; it can only accelerate to 30–40 per minute in diffuse affections of fine bronchi and bronchioles. Dyspnoea and tachycardia develop.

Percussion sounds over the lungs are usually unchanged. Ausculation reveals harsh breathing and dry buzzing and whistling rales, which often change their character and amount after cough. During resolution of inflaammation in the bronchi, tenacious sputum is thinned by the action of proteolytic enzymes contained in it, and moist dulled rales may be heart together with dry rales. X-ray examination in acute bronchitis does not reveal any change. The lelucocyte count of the blood may rise to reveal any change. The leucocyte count of the blood may rise to reveal any change. The leucocyte count of the blood may rise to 9000–11000 in one microlitre. ESR slightly increases. The sputum is mucous or mucopurulent, sometimes with streaks of blood; it contains columnar epithelium and other cell elements. Fibrin clots (bronchial casts) are expectorated in acute fibrous bronchitis.

Course

The patient usually recovers in two or three weeks. Under the effect of some aggravating factors, such as smoking or chilling, or in the absence of timely treatment, the disease may run a aprotracted course or become complicated with bronchopneumonia.

Treatment

Antibiotics and sulphonamides, broncholytics (ephedrine or

isoprenaline) in combination with expectorants, e.g. thermopsis influsion (1:200), a tablespoonful three times a day, alkaline inhalations and othe remedies are prescribed. Antibiotics and broncholytics should better be taken as aerosols. Also recommended are cups, mustard plasters and compresses on the chest, foot baths.

Prophylaxis

Hardening of the body (cool showers, bathing); removal of irritants of the upper air ways (smoking, occupational hazards) and treatment and prevention of chronic diseases of the nose and throat.

Chronic Bronchitis

Chronic bronchitis is chronic inflammation of the bronchi and bronchioles.

Aetiology and pathogenesis

Infection is an important factor in development and further rcourse of the disease. Chronic bronchitis may develop after acute bronchitis or pneumonia. Chronic occupational inhalation of dusts or chemicals, and also residence in cities with moist climate and frequent fluctuations of the weather conditions, are also predisposing factors. No less important factor for the development of chronic bronchitis is smoking: bronchitis occurs in 50–80 per cent of habitual smokers while its incidence among non-smoking population is only 7-19 per cent. Autoimmune allergic reactions occurring due to absorptionof products of protein decomposition (that are formed in the lungs in inflammatory foci) are important factors for persisten tchronic bronchitis.

Pathological anatomy

At the early stage of the disease, the mucosa is plethoric, cyanotic, and hypertrophic at some points. Mucous glands are hyperplastic. Inflammation eventually involves the submucous and muscle layers, where cicatricial tissue develops. The mucosa and cartilaginous lamina become artrophied. Bronchial walls become thin to increase the lumen and to cause bronchiectasis. The peribronchial tissue can also be involved in the process with subsequent development of interstitial pneumonia. Interalveolar septa become gradually atrophied and lung emphysema develops; the number of capillaries of the pulmonary artery decreases. Muscular hypertrophy of the right ventricle and also right-ventricular incompetence may join the pulmonary insufficiency.

Clinical picture

The picture of chronic bronchitis depends on the degree of bronchial involvement and on the depth of affection of the bronchial wall. The main symptoms of chronic bronchitis are cough and dyspnoea. Cough may differ in character and change depending on the season and weather. In dry weather, especially in summer, cough is mild and dry, or it may be absent. At increased air humidity and in rain, cough intensifies, while in the cold season it becomes persistent and intense, with expectoration of tenacious mucopurulent or purulent sputum. Sputum is sometimes so thick that it is expectorated in the form of fibrinous bands that lood like casts of the bronchial laumen (fibrinous bronchitis).

Dyspnoea in chronic bronchitis is not only due to the lung hypoventilation but also due to lung emphysema that develops as a secondary process. The character of dyspnoea is mostly mixed. First difficult breathing occurs only during exercise such as ascending a hill or upstairs. Further dyspnoea becomes more pronounced. In diffuse inflammation of fine bronchi, dyspnoea becomes expiratory. The disease may be attended by the general symptoms such as weakness, rapid fatigue excess perspiration; temperatuer rises in exacerbation of bronchitis.

Inspection palpation and percussion of the chest, and also X-ray examination do not reveal any changes in non-complicated chronic bronchitis. In grave chronic bronchitis, when pneumosclerosis, pulmonary emphysema, and cardiopulmonary insufficiency attend the main process, accessory muscles become actively involved in the respiratory act, the neck veins become swollen, and cyanosis develops. In this case percussion of the lungs gives a bandbox sound; mobility of the lower border of the lung is limited. Breathing can be vesicular, harsh, or (when emphysema develops) weakened vesicular. Buzzing and whistling dry (less frequently moist) rales are also heard on ausculation.

The blood picture only changes in exacerbation of bronchitis. The change include increased leucocyte counts and acclerated ESR.

Sputum of a patient with chronic bronchitis is *mucopurulent* or *purulent*. In putrefactive bronchitis, sputum is greenish-brown or brown because of decomposed blood; the smell is offensive. Microscopy reveals a great number of ieucocyes, degrading erythrocytes, and ample coccal flora.

X-ray examination in chronic bronchitis, complicated by developing pneumosclerosis or lung *emphysema*, reveal signs of these diseases. Deformation of the bronchi may be revealed on bronchography. Bronchoscopy gives a picture of atrophic less frequently hypertrophic) bronchitis (with thinning or swelling of the bronchial mucosa).

Course

The course of chronic bronchitis varies. It may persist for many years, but the signs of the anatomical and functional changes are mildly pronounced. In other patients chronic bronchitis progresses slowly to give exacerbations on chills and during epidemic outbreaks of influenza, or under the effect of some occupational hazards. Recurrent bronchitis and peribronchitis give bronchiectases and often pneumonia. Obstruction of bronchial patency provokes the development of emphysema and cardiopulmonary insufficiency.

Treatment

In chronic *bronchitis* treatment is aimed at eradication and subsequent prevention of exacerbations (the patient should be kept under regular medical observation). The inflammatory process is the bronchi should be removed. In aggravation of chronic bronchitis sulpha drugs tibiotics are prescribed. If the disease is protracted and aggravated by asthma, broncholytics are given in addition. Corticosteroids, which have anti-allergic and anti-inflammatory action, are given in some cases. To lessen cough, antitussives (codeine and libexin) and expectorants (thermopsis infusion, 1.0 : 200.0, a tablespoonful 3-4 times a day) are prescribed. Harmful occupational and domestic factors (smoking in particulars) should be removed

Climate therapy favours recovery

Warm marine climate is especially helpful.

Prophylaxis consists is removal of harmful factors provoking the onset of inflammation of the bronchi (e.g. smoking), improvement of working conditions (especially in industry where air may be contaminated with dusts, etc.), treatment of chronic diseases of the upper airways, and strengthening the organism's defence forces (hardening by cool showers, swimming, exercises, walks, etc.).

Bronchial Asthma

Bronchial asthma is an allergic disease which is mainifested by paroxysmal attacks (Gk asthma panting).

Aetiology and pathogenesis

Bronchial asthma is a polyaetiological disease. It can be provoked by external agents (exogenic allergens) and internal causes (endogenic allergens that usually depend on the infections of the airways). Non-infectious allergic (atopic) and infectious-allergic asthma are distinguished accordingly.

Attacks of asthma can be provoked by various odours, such as those of flowers, hay, perfumes, petrol, carpet or pillow dusts, moulds (in damp rooms), the smell of urosol dyes, asbestos dust; by some foods such as eggs, crabs, strawberries (food asthma), and medicinal preparations. An attack of asthma can sometimes be provoked not by the allergen itself but by memory of it or by remembrance of the conditions under which the allergen acted in the past. The patient can develop an asthmatic attack when he reappears in a certain room or a house or street, where he once had an attack, or oven by remembrance of this particular room, house, street, etc.

Endogenic allergens causing attacks of asthma include microbial antigens that are formed during various inflammatory processes, such as sinusitis, chronic bronchitis, chronic pneumonia, etc. Products of decomposition of microbes and tissue proteins forming due to proteolytic process at the inflammatory focus can act as allergens.

Asthma is an alergic disease in which the body is sensitized mainly by the substances of protein nature; other substances and effects can also provoke the disease.

The allergen becomes active in the body only under certain conditions. Bronchial asthma andits attacks develop when the body's reactivity is changed. An important factor promoting sensitization of the body and pathological reactivity is the hereditary-constitutional factor. Climate is also important in this respect. For example, attacks of asthma more often occur during spring or autumn; they can be removed in certain climatic zones, e.g. in highland conditions, or, on th contrary, they may be intensified. Attacks are more likely to occur in cold and damp weather. Pathological response of the parasympathetic nervous system to stimuli received from various exteroceptos and interoceptors, hyperexcitation of the vagus nerve centres, development of pathological reactions in the afferent receptors of the bronchial walls and their hypersensitivity to local irritants are also important factors in the development of asthma attacks. There is evidence of the important factors in the development of asthma attacks. There is evidence of the importance of partial b-adrenergic blockade in the pathogenesis of bronchial asthma and reduced activity of cyclic adenosine monophosphate (intracellular mediator) during attacks of the disease. Finally, certain hormonal shifts are also important, which in the first instance is connected with the adrenal glands. This is confirmed, in particular, by the favourable effect of corticosteroid therapy on the course of the disease.

It has recently been established that attacks of asthma are provoked by an allergic reaction occuring in the bronchial tissue. Three stages in the course of an attack of atopic bronchial asthma, which proceeds as an allergic reaction of the immediate type, are distinguished. In the first (immunological) stage the antigen combines with the specific antibodies reagins (the IgE class compounds fixed mainly on mast, plasma, and lymph cells); in the second (pathochemical) stage histamine, serotonin, a slow-acting substance anaphylactin, and other biologically active substances are released from these cells as a result of degranulation and alteration; in the third (pathophysiological) stage, spasms develop in the bronchi along with the exudative reaction and oedema of the bronchial mucosa, as the result of the action of the above mentioned substances. An attack of an infectious-allergic asthma develops in the same way, but a delayed (Cellular) allergic reaction may be involved in its origination.

Clinical picture

The classical description of bronchial asthma was given in 1838 by G. I. Sokolsky. An attack of allergic asthma begins abruptly and acutely and usually quickly subsids. Attacks of dyspnoea developing against the background of chronic infectious diseases of the respiratory ducts (infectious-allergic asthma) are often not severe but protracted. Signs of chronic bronchitis, pneumosclerosis, and lung emphysema can be revealed in such patients in periods clear of paroxysms.

Attacks of dyspnoea in bronchial asthma are quite similar; they arise suddenly, gradually increase in strength, and last from a few minutes to several hours and even several days. A prolonged attack of asthma is called status asthmaticus. During such an attack, the patient has to assume a forced attitude; he usually sits in bed, leans against his laps, his breath is loud, often whistling and noisy, the mouth is open, the nostrils flare out. The veins of the neck become swollen during expiration and return to norm during inspiration. At the peak of an attack, the patient begins coughing with poorly expectorated thick and tenacious sputum. The chest expands during an attack (to the size of the chest during inspiration). Accessory respiratory muscles are actively involved in the respiratory act. Percussion of the lungs gives the bandbox sound, the lower margins of the lungs are below normal, the mobility of the lower borders is sharply limited during both inspiration and expiration. Ausculation reveals many whistling rales against the background of weakened vesicular respiration with a markedly prolonged expiration. The whistling rales are sometimes heard even at a distance. Tachycardia is usually observed. The borders of complete dullness of the heart cannot be determined because of the acute inflation of the lungs. By the moment the attacks abates the sputum thins and expectoration becomes easier: high and dry rales in the lungs determined by ausculation decrease to give ways to low buzzing and often moist non-consonant rales of various calibres; the attack of dyspnoea gradually abates.

Blood test during attacks shows moderate lymphocytosis and consinophilia. The sputum contains 40 to 60 per cent of eosinophils and often Curschmann spirals and Charcot-Leyden crystals.

X-ray examination of the thoracic organs during an attack of asthma shows high translucency of the lung fields and limited mobility of the diaphragm.

Patients with uncomplicated bronchial asthma have no complaints in the periods clear to attacks. Physical, X-ray, and laboratory examinations reveal no changes except eosinophilia of the blood.

Course

Attacks of asthma sometimes occur very rarely (once a year or even several years). Some patients develop a more severe course with frequent and grave attacks. Concurrent chronic bronchitis, pneumosclerosis, and emphysema of the lungs cause the corresponding changes detectable by routine examinations; cardiopulmonary insufficiency gradually develops. In rare case the patient may die during an attack.

Treatment

The provoking factor should be first identified and eliminated whenever possible. The removal of the causative stimulus will prevent the development of the disease. The allergen can be identified by sensitivity skin tests using suspected substances in special allergological laboratories. Once the allergen is known, desensitization should be attempted. Foci of infection should be treated in infectious-allergic asthma (bronchitis, bronchiectasis, sinusitis, etc.). Unfortunately it is not always easy to remove the cause of bronchial asthma or to carry out effective pathogenic therapy. Symptomatic therapy should then be given to remove or prevent attacks of asthma.

An attack of asthma can be removed by subcutaneous injections of 0.2–1 ml of a 0.1 per cent adrenaline hydrochloride solution or 1 ml of a 5 per cent ephedrine solution. Euphylline (aminophylline) is given intravenously. Broncholytics are also prescribed (inhalations included). A protracted status asthmaticus is an indication to corticosteroid therapy. To prevent attacks of bronchial asthma, broncholytics are given in courses. Remedial exercise and health-resort therapy are also indicated.

ACUTE PNEUMONIA

Pneumonia is an acute inflammation of the lungs developing either independently or as a complication in other diseases. Pneumonia is prominent among other diseases of the internal organs. It prevalis in epidemic influenza. Men are more susceptible to the disease than women; it is especially severe in children and elderly patients.

An aetiological classification of pneumonia has been adopted. Differentiated are bacterial pneumonia (pneumococcal, staphyloccal, streptococcal, etc.), virus (caused by the influenza virus, viruses or ornithosis, psittacosis), mycotic (candidiasis, etc.) and pneumonia caused by irritating gases, vapours, dusts, etc. it has long been established that the clinical aspects of bacterial pneumonia somewhat vary depending on the initial body's reactivity. This mainly refers to pneumonia caused by pneumococcus, especially of types I and II. Sometimes the disease is hyperergic, which accounts for special acuity, cyclic character of its course, frequent affection of the whole lobe of the lung (with involvement of the pleura) nd a special character of effusiondue to a markedly impaired permeability of the vessel wall (the presence of fibrin and erythrocytes in the effusion). As distinct from more frequently occurring bronchopneumonia (syn: focal, catarrhal,l lobular pneumonia), this pneumonia is called croupous (syn.: acute lobar, fibrinous pneumonia, pleuropneumonia). Pleuropneumonia was formerly diagnosed very frequently but now it rarely occurs in its typical form.

Bronchopneumonia (Focal Pneumonia)

Separate lobules of the lungs are affected in bronchopneumonia, hence another name, lobular pneumonia. Inflammatory foci may be multiple, or they may fuse (confluent pneumonia); the foci may be located in various parts of both lungs simultaneously (mostly in the lower parts of the lungs).

Aetiology and pathogenesis

Quite varied bacterial flora would be normally found in bronchopneumonia. The wide use of antibiotics has changed the proportion of microbes that ar found in pneumonia. The importance of pneumococci has significantly decreased while the role of other microorganisms, especially of streptococci and staphylococci, has increased. Acute pneumonia is caused in many cases by viruses (in influenza, ornithosis, and psittacosis).

In addition to the infectious factor, predisposing conditions are also very important. They decrease the immunological properties of the body (overcooling, acute respiratory diseases, etc.). Bronchopneumonia can develop against the background of chronic disease of the lungs (bronchiectasis chronic bronchitis) due to haematogenic infection in purulent inflammatory diseases (sepsis, after operations, etc.) Aged patients with long-standing and severe diseases or subjects with plethoric congestion of the lungs can develop hypostatic pneumonia. Aspiration of foreign bodies (food, vomitus, etc.) causes aspirations pneumonia. Inhalation of suffocating or irritating gases or vapours (benzene, toluene, benzine, etc.) or other toxic substances can also provoke the onset to bronchopneumonia. The aetiological factor is often decisive for the clinic and couse of pneumonia, but irrespective of aetiology, there are always some general signs of the disease.

Development of bronchopneumonia is associated with the extension of the inflammatory process from the bronchi and bronchioles to the pulmonary tissue (hence another name of bronchopneumonia–catarrhal pneumonia, which reflects the transition of inflammation and infection with the mucous secretion from the inflamed bronchi into the alveoli). Infection gets inside the pulmonary tissue via the bronchi, and more frequently peribronchially, i.e. by lymph ducts and interalveolar septa. Local atelectasis that occurs in obstructionof the bronchus by a "mucopurulent plung" (N. Filatov) is important in the pathogenesis of bronchopneumonia. Obstruction of bronchial patency can be caused by a sudden bronchospasm an doedema of the bronchial mucosa, inflammation (bronchitis), etc. Recently bronchopneumonia occurs mostly in children and the aged, usually during cold seasons (spring, autumn, winter).

Pathology anatomy

The microscopic picture of inflammatory foci in bronchopneumonia is quite varied because of the different persistence of the foci. The alveoli at the site of inflammation are filled with serous or mucous effusion containing large amount of leucocytes. If bronchopneumonia is associated with influenza, microscopy shows the rupture of fine vessels. In confluent pneumonia, the inflammatory foci fuse together to involve several segments or even the whole lobe of the lung.

Clinical picture

The onset of the disease is usually overlooked because it often develops against the background of bronchitis or catarrh of the upper airways. The findings of physical examination of the patient at the onset of bronchopneumonia are the same as in acute bronchitis. Sites of consolidated tissue are often small and difficult to reveal by X-ray examination. But if a patient with clinical signs of acute bronchitis develops high temperature and has symptoms of a more severe disease, he should be considered to have bronchopneumonia.

Themost typical signs of bronchopneumonia are cough, fever, and dyspnoea. If the inflammatory focus is at the periphery of the lung and the inflammation involves the pleura, pain in the chest during coughing and deep breathing may occur. Fever may persist for various terms in bronchopneumonia. Usually fever is remittent and irregular. The temperature is often subfebrile or it may even be normal in the middle-aged or old patients.

Objective examination can sometimes reveal moderate hyperaemia of the face and cyanosis of the lips. Respiration accelerates to 25–30 per min; respiratory lagging of the affected side of the chest may be observed. Percussion and ausculation may prove ineffective if the inflammatory foci are small and deeply located. In the presence of a large focus, especially if it is located at the periphery of the lung tissue, and also in confluent pneumonia the percussion sounds lose resonance (or become completely dull), and ausultation reveals vesiculobronchial or bronchial breathing. Vocal fremitus and bronchophony are characteristic of such cases. Dry and moist rales are frequent, but consonating moist rales and crepitation that are heard over a limited part of the chest are especially informative. X-ray examination reveals indistinct densities; in confluent pneumonia densities are spotted (mostly in the lower portions of the lungs). The shadows of the lung roots may be expanded due to enlarged lymph nodes. X-ray examination of the lungs reveals focal inflammations at least 1–2 cm in diameter; very small and separated foci of consolidabed lung tissue are undeterminable; therefore in the absence of the X-ray signs of pneumonia the diagnosis of bronchopneumonia cannot be rejected in the presence of clinical symptoms.

Sputum is mucopurulent, first tenacious but later more thin, sometimes there are traces of blood, but it is not rusty. It contains a great number of leucocytes, macrophages and columnar epithelium. Bacterial flora is varied and ample. But it is scant in virus pneumonia. The blood count shows mild neutrphilic leucocytosis, a certain shift to the left, and a moderately increased ESR.

Course

Bronchopneumonia is usually more protracted and flaccid than pleuropneumonia. Prognosis is favourble with appropriate treatment. But the disease can transform into its chronic form. Bronchopneumonia can be aggravated by abscess of the lung and bronchiectasis.

Acute Lobar Pneumonia

Aetiology and pathogenesis. All authors who studied the aetiology of acute lobar pneumonia (pleuropneumonia, crupous pneumonia), discovered Frenkel pneumococci (mostly types I and II, less frequently types III and IV) in about 95 per cent of cases. Fridlaender diplobacillus, Pfeiffer's bacillus, streptococcus, staphylococcus, etc. ar found less frequently.

Acute lobar pneumonia occurs mostly after severe overcooling. The main portal of infection is bronchogenic, less frequently lymphogenic and haematogenic. Congestion in the lungs in cardiac failure, chronic and acute diseases of the upper airways, avitaminosis, overstrain and other factors promote the onset of pneumonia. Acute lobar pneumonia is relatively frequent in patients who had pneumonia in their past history (it recurs in 30–40 per cent of cases which is another evidence of the hypergic character of the disease).

Clinical Picture

The onset of the disease. Typical acute lobar pneumonia begins abruptyl with shaking chills, severe headache, and fever (to 39–40°C). The chills usually persist for 1–3 hours, then pain appears in the affected side; sometimes it may arise below the costal archin the abdomen to simulate acute appendicitis, hepatic colics, etc. (this usually occurs in inflammation of the lower lobe of the lung, when the diaphragmal pleura becomes involved in the process). Cough is first dry and in 1-2 days rusty sputum is expectorated. The patient's general condition is grave. General examination shows hperaemia of the checks, more pronounced on the affected side, dyspnoea, cyanosis, often herpes on the lips and nose; the affected side of the chest lags behind in the respiratory act. Vocal fermitus is slightly exaggerated over the affected lobe. Sounds over the lungs are quite varied and depend on the distribution of the process, the stage of the disease, and other factors. At the onset of the disease, shortened percussion sound can be heart over the affected lobe, often with tympanic effect because liquid and air are simultaneously contained in the alveoli; the vesicular breathing is decreased while bronchophony is increased; the so-called initial crepitation (crepitus indux) is present.

The height of the disease (classified by pathologists as the red and grey hepatization stages) is characterized by the grave general condition. It can be explained not only by the size of the affected area of the lung which thus does not take part in respiration but also by general toxicosis. Respiration is acclerated and superficial (30–40 per min) and tachycardia (100–200 beats per min) is characteristic. Dullness is heart over the affected lobe of the lung; bronchial respiration is revealed by ausculation; vocal fremitus and bronchophony are exaggerated. Vocal fremitus is in some cases either absent or enfeebled (in combination with pleurisy with effusion, and also in massive acute lobar pneumonia, in which the inflammatory exudate fills large bronchi); bronchial breathing is inaudible. Before the antibiotic era, the patient with acute lobar pneumonia would often develop vascular failure with a marked drop in the arterial pressure due to toxicosis. Vascular collapse is attended by general asthenia, drop of temperature, increased dyspnoea, cyanosis, and accelerated and small pulse. The nervous system is also affected (sleep is deranged, hallucinations and delirium are possible, especially in alcoholic patients). The heart, liver, kidneys and some other organs are also affected. Fever persists for 9–11 days if antibiotics or sulpha drugs are not given. The temperature then drops either abruptly durign 12–24 hours or lytically, during more than 2–3 days.

Resolution stage

The exudate thins, air again fills the alveoli to decrease dullness of the percussion sound, tympany increases and bronchial breathing lessens. Crepitation is heard again (crepitus redux) because the alveolar walls separate as air fills them. Moist rales are heard. Exaggerated vocal fermitus, then bronchophony, and finally bronchila breathing disappear. The leucocyte count in the blood increases to 15 × 109–25 × 109 per liter (15000–25000 per microlitre); neutrophils account for 80-90 per cent of the leucocytes; a shift to the left with the appearance of juvenile forms is sometimes observed. The number of eosinophils decreases and they can disappear completely in grave cases. Relative lymphopenia and monocytosis are observed. The ESR increases. The red blood does not change.

Sputum is tenacious during the congestion period; it is slightly crimson and contains much protein, a small number of leucocytes, erythrocytes, alveolar cells, and macrophages. In the stage of red hepatization sputum is scant and rusty; it contains fibrin and a higher number of formed elements. In the stage of grey hepatization leucocyte count in the sputum increases significantly; the sputum becomes mucopurulent. In the resolution stage, leucocyes are converted into detritus, which is found in the sputum; many marcrphages are also found. Pneumococci, staphylococci, Friedlaender diplobacilli can be detected in the sputum.

X-ray changes in the lungs depend on the stage of the disease. The lung patterns is first intensified, then dense foci develop, which later fuse. The shadow usually corresponds to the lung lobe. The lungs becomes normally clear in two or three weaks. Dynamics of the X-ray changes depends on the tiem when the therapy is begun.

Course and complications. Fatal outcomes in acute lobar pneumonia were formely 20-25 per cent. Before antibiotics and sulpha drugs came into wide use, compolications (purulent processes in the lungs and pleurisy, usually purulent) wer frequent; their incidence has now decreased significantly. If pleurisy develops before resoltuion of pneumonia, it is called parapneumonic. If it occurs after resolution, it is referred to as metapneumonic pleurisy. Other complications occur as well: myocarditis, meningitis, and focal nephritis. If resolution of the exudate is delayed, and connective tissue grows into it, cirrhosis of the affected lobe of the lung develops (carnification). Mortality from acute lobar pneumonia has recently decreased significantly.

Treatment of pneumonia

The patient should be hospitalized and kept in bed. Food should be rich in viamins and easily assimilable. Antibiotics (penicillin, 300 000 - 500 000 units, 4 to 6 times a day intramuscularly, streptomycin, tetracycline and other antibiotics) and supha drugs (sulphadimezine, sulphaethiodole, sulphadimethoxine, etc.) are the main preparations for treating pneumonia. Oxygen therapy (an oxygen mask or a tent) is also helpful. It improves metabolism and has a favourable effect on the cardiovascular system. Vasicular insufficiency is treated by coffein and camphor; in the presence of heart failure, digitalis and strophanthine preparations are given. Expectorants (thermopsis, etc.) should be given, especially during the resolution of pneumonia; cups mustard plaster, and physiotherapy accelerate elimination of residual effects of pneumonia. Respiratory exercises are important to improve lung ventilation.

Prophylaxis mainly consists in strengthening and hardening of the body.

PULMONARY ABSCESS

Pulmonary abscess is a purulent melting of the lung tissue circumscribed by an inflammatory swelling.

Aetiology and pathogenesis

The aetiological factor in the development of a purulent process in the lungs is mostly the coccal flora (streptococci, staphylococci, and pneumococci). The disease is sometimes associated with saprophytic autoinfection of the upper airways. The development and intensity of the purulent process in the lungs depend on the reactivity of the lung tissue and of the entire body.

The purulent process in the lungs develops mostly as an outcome of pneumonia or complicated bronchiectasis. Primary abscesses of the lungs arise in wounds to the chest, aspiration of foreign bodies, and after operations on the upper airways (tonsillectomy and the like). Pulmonary abscess can develop also by a haematogenic or lymphogenic routes, when the infection is carried into the lungs from a purulent focus in the body. The lung tissue can be resorbed without the involvement of microbes. In lung infarction, or in the presence of a degrading tumour, a cavity may be formed due to local disorder in blood circulation. In such cases necrotization of the lung tissue can be the precursor of a purulent process. Suppuration of an echinococcal cyst in the lung can be another cause.

Clinical picture

Two periods are distinguished: before and after opening of an abscess. The first period is formation of the abscess. It continues for 10-12 days, on the average. At the onset of the sisease patients complain of general indisposition, weakness, chills, cough with mearge sputum, and pain in the chest. Fever is first moderate but gradually becomes remittent, and then hectic. Dyspnoea develops even in the presence of a small abscess.

Palpation of the chest can in some cases detect painfulness in the intercostal spaces of the affected side. This symptom is associated with involvement in the process of the costal pleura. Unilateral thoracic lagging corresponding to inflammation may be observed. Vocal fremitus depend on the location of the inflammatory focus. It is often exaggerated in peripheral location and remains unchanged in deep location of the inflammatory focus.

Percussion can determine loss of resonance (or dull sound) on the affected side. Ausculation in small deep abscesses does not reveal any abnormalities. If the abscess is superficial, respiration over the affected side is decreased and vesicular, sometimes with a bronchial character. In some cases breathing is harsh, and dry rales can be heard.

The blood picture is an important and objective criterion in the diagnosis of lung abscess. Neutrophilic leucocytosis, 15 × 10-9 – 25 × 109 per litre (15000–20000 per microlitre), is observed; a shift to the left, to the myelocytes, is also characteristic. The ESR increases significantly. Study of the sputum is not specific before the abscess opens.

The X-ray picture of the abscess during the first period of the disease does not differ from that in common pneumonia or tuberculous infiltration; a large focus of increased density with rough and indistinct margins is determined.

The clinical picture of the second period begins with the opening of the purulent abscess into the bronchus. There may be a transitory condition between the first and second periods of the disease during which the main clinical symptoms are intensified (elevated temperature with great circadian variations, heavy cough, dyspnoea, pain in the chest, etc.). Rupture of the abscess into the bronchus is attended by a sudden release of ample purulent sputum ("full mouth"), sometimes with offensive odour, which on standing separates into two or three layers: mucous, serous, and purulent. Further expectorations will depend on the volume of the cavity and vary from 200 ml to 1-21 a day. Objectively the patient looks feverish. If the pleura is involved, unilateral thoracic lagging is observed on the affected side. After the cavity is free from pus, and depending on its location and size, physically findings may be different. Tympany is determined in the presence of large and superficial abscesses by percussion. Respiration can be either bronchovesicular or bronchial; if a large cavity containing air communicates with the bronchial lumen, respiration can be amphorical. Resonant moist moderate and large bubbling rales are usually heard over a limited area. In the presence of concurrent diffuse bronchitis, intense rales interfere with location on the abscess.

Neutrophilic leucocytosis with a shift to the left and increased ESR are observed. In grave and protracted cases, iron-deficiency (hypoferric) anaemia develops.

On standing, sputum separates into three layers: the upper (foamy and mucous), the middle (liquid, serous) and the bottom layer (pus). Microscopy of the sputum reveals the presence of elastic fibres (in addition to many leucocytes and erythrocytes). The absence of elastic fibres in the sputum indicates termination of degradation of the lung tissue. Bacteriological flora is ample, mostly containing cocci; in the presence of antibiotic therapya the flora is meagre.

X-raying of the emptied abscess shows a specifically increased translucency. Variation of the liquid level can be observed with variation of the patient's posture. If the draining bronchus is at the bottom of the cavity (which is common with abscesses of the lung apices), the liquid level is undeterminable (the fluid flows down and is withdrawn through the outlet bronchus). The abscess cavity is surrounded on all sides by a border of inflamed tissue with a diffuse outer contour.

Course

Development of the lung abscess and its healing depend on location of the cavity, conditions for its emptying, and concurrent complications. In most cases abscesses heal with formation of focal pulmonary fibrosis at the site of the common cavity, but in 30-40 per cent of patients recovery is incomplete and the diseae becomes chronic. Chronic diseases of the lungs (bronchiectasis pulmonary fibrosis, etc.) promote a protracted course of pulmonary suppuration. The following complications of the lung abscess are distinguished: (1) rupture of the abscess into the pleural cavity with development of pyopneumothorax; (2) pulmonary haemorrhage; (3) development of new pulmonary abscess; (4) metastases of the abscesses into the brain, liver and other organs. Complications aggravate the prognosis of the disease.

Treatment

Bed-rest regimen in hospital is necessary. Antibiotics and sulpha drugs should be given as early as possible. Penicillin and streptomycin should be given along with broad-spectrum antibiotics such as tetracycline (0.2-0.3 g 4-6 times a day after meals, oletetrin, etc.). Sulpha drugs (sulphadimidine, sulphadimethoxine, etc.) should be given in sufficiently large doses and for a long period. In addition to peroral or intramuscular administration, antibiotics should be given intratracheally or as aerosols (1–2 times a day for 2–4 weeks). Therapeutic bronchoscopy is widely used to remove pus from the cavity and to administer antibiotics directly into it. Symptomatic treatment consists in prescribing expectorants and broncholytics (ephedrine, drotaverine, and others) which liquefy sputum. Adequate draining is very important for a better empthing of the cavity: the the patient should find a posture at which sputum withdrawal is the best and to assume this posture for 30 minutes two or three times a day.

If conservative therapy proves ineffective, surgical treatment should be given in a month or two (earlier, in gravae cases). The operation consists in removal of the lobe (or the whole lung if abscesses are multiple).

PLEURISY

Pleurisy is inflammation of the pleura. Dry pleurisy (pleuritis sicca) and pleurisy with effusion (pleuritis exudativa) are distinguished. The character of the inflammatory effusion may be different: serous, serofibrinous, purulent, and haemorrhagic.

Aetiology and Pathogenesis

Serous and serofibrinous pleurisy attend tuberculosis (in 70-90 percent of cases), and pneumonia, certain infections, and also rheumatism in 10–30 per cent of cases. The purulent process in the pleura may be caused by pneumococci, streptococci, staphylococci, and other microbes. Haemorrhagic pleurisy arises in tuberculosis of the pleura, bronchogenic cancer of the lung with involvement of the pleura, and also in injuries to the chest.

Most diseases of the pleura (pleurisy included) are secondary to disease of the lung. Pleurisy usually develops as a reaction of the pleura to pathological changes in the adjacent organs, in the lungs in the first instance, and less frequently as a symptom of a systemic disease (polyserosites of various aetiology). Serous pleurisy often arises as an allergic reaction. Purulent pleurisy is often a complication of bronchopneumonia: inflammation may extend onto the pleura, or an inflammatory focus may turn into an abscess which opens into the pleural cavity. Inflammation of the pleura is always attended by markedly increased permeability of the wall of the affected capillaries of the pulmonary pleura.

Reactivity of the body is a very important factor in the pathogenesis of pleurisy. In fibrinous or dry pleurisy fibrin precipitates from the exudate (which is produced in a small amount) and gradually deposits on the pleura. Serous pleurisy may become infected to convert into purulent; exudate becomes turbid and contaisn many leucocytes. In the presence of purulent processes in thelungs or adjacent organs (pericarditis, perioesophagitis, etc.), purulent pleurisy often develops abruptly. The affection of the pleura in tumours, which in most cases are metastatic (less frequently primary), decreases its absorptive function to promote accumulation of pleural effusion (haemorrhagic effusion in most cases).

Dry Pleurisy

Clinical picture

A characteristic symptom of dry pleurisy is pain in the chest which becomes stronger during breathing and coughing. Cough is usually dry, the patient complains of general indisosition; the temperature is subfebrile. Respiration is superficial (deep breathing intensifies friction of the pleural membranes to cause pain). Lying on the affected side lessens the pain. Inspection of the patient can reveal unilateral thoracic lagging during respiration. Percussion fails to detect any changes except decreased mobility of the lung border on the affected side. Ausculation determines pleural friction sound over the inflamed site. X-ray picture shows limited mobility of the diaphragm because the patient spares the affected side of his chest.

The blood picture remains unchanged but moderate leucocytosis is observed in some cases.

Course

Dry pleurisy has a favourable course and the patient recovers completely in one or three weeks.

Pleurisy with Effusion

Clinical picture

Patients suffering from pleurisy with effusion usually complain of fever, pain or the feeling of heaviness in the side, and dyspnoea (which develops due to respiratory insufficiency caused by compression of the lung). Cough is usually mild (or absent in some cases). The patient's general condidtion is grave, especially in purulent pleurisy, which is attended by high temperature with pronounced circadian fluctuations, chills, and signs of general toxicosis. Inspection of the patient reveals asymmetry of the chest due to enlargement of the side where the effusion is accumulated; the affected side of the chest usually lags behind in respiratory movements. Vocal fermitus is not transmitted at the area of fluid accumulation.

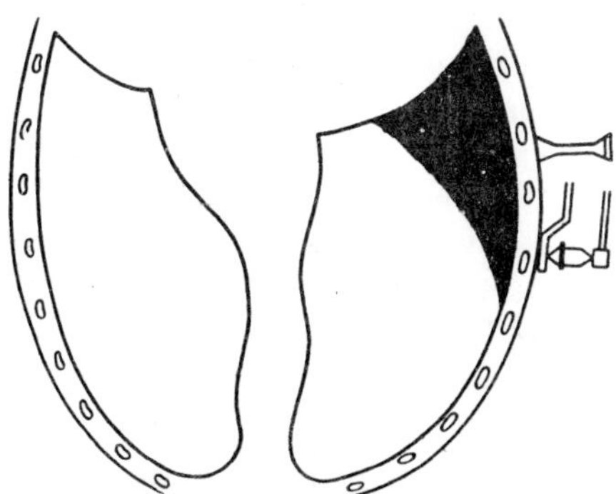

Fig. 1.1. Percussion and auscultation in pleurisy with effusion.

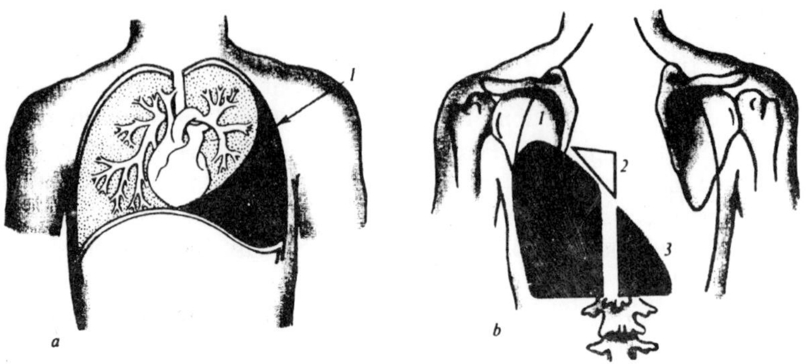

Fig. 1.2. Pleurisy with effusion.
a—anterior view; b—posterior view; 1—Damoiseau's curve; 2—Garland's triangle; 3—Rauchfuss-Grocco triangle.

Percussion over the area of fluid accumulation produces dullness. The upper limit of dullness is usually the S-shaped curve (Damoiseau's curve) whose upper point is in the posterior axillary line. The effusion thus occupies the area, which is a triangle both anteriorly and posteriorly. The Damoiseau curve is formed because oxudate in pleurisy with effusion more freely accumulates in the lateral portions of the pleural cavity, mostly in the costal-diaphpragmatic sinus. As distinct from effusion, which is restricted by adhesions, tranudate more freely presses the lung and the Damoiseau curve is not therefore determined. In addition to th eDamoiseau curve, two triangles can be deermined by percussion in pleurisy with effusion. The Garland triangle is found on the affected side and is characterized by a dulled tympanic sound. It corresponds to the lung pressed by the effusion, and is located between the spine and the Damoiseau curve. The Rauchfuss-Grocco triangle is found on the healty side and is a kind of extension of dullness determined on the affected side. The sides of the triangle are formed by the diaphragm and the spine, while the continued Damoiseau curve is the hypotenuse. The triangle is mainly due to displacement of the mediastinum to the healthy side. Mobility of the lower border of the lung on the affected side is not usually determined in pleurisy with effusion. Left-sided pleurisy with effusion is characterized by the absence of the Traube space (the left pleural sinus is filled with effusion and a dulled percussion sound is heart over the gastric air bubble instead of the tympany).

Respiration in the region of accumulated effusion is not ausculated, or it can be very weak. Respiration ausculated slightly above the effusion level is usually bronchial which is due to compression of the lung and displacement of air from it. Vocal fremitus and bronchophony over the effusion are not determined because the vibrating walls of the bronchi that conduct voice are separated from the chest wall by the fluid. The heart is usually displaced by the effusion toward the healthy side. Tachycardia is observed. Arterial pressure may be decreased. Dizziness, faints, etc., sometimes occur because of the marked toxicosis.

A radiograph of the thoracic organs shows a homogenous density whose area corresponds to the area of dullness. If effusion is scarce, it accumulates in the outer sinus. Large volumes of effusion cover the entire lung to its apex and displace the mediastinum toward the intact side to lower the diaphragm. The encapsulated parietal pleurisy gives the picture of parietal density. The medial border is usually sharply outlined. Density of the interlobar pleurisy extends along the interlobar sulcus in the form of a triangle or a spindle. Diaphragmatic pleurisy is charactrized either by a limited mobility of the diaphragm or its complete absence. The upper border of the effusion is convex (upward) to follow the curvature of the diaphragm.

Exploratory puncture is necessary in order to determine the properties of the exudate for accurate diagnosis. The fluid obtained by pleurl puncture is studied at the laboratory.

During the initial stage of the disease the blood picture may show mild leucocytosis (marked leucocytosis is characterstic of purulent pleurisy) and sometimes eosinophilia. The ESR is increased. Tuberculosis pleurisy is characterized by lymphocytosis, while rheumatic pleurisy by neutrophilosis.

Course

The course of pleurisy with effusion depends on the aetiology of the affection. Pleurisy in rheumatism would normally resolve in 2–3 weeks (with appropriate treatment). Pleurisy with effusion complicating pneumonia (metapneumonic pleurisy, usually serous) also has a comparatively mild course. A protracted course is characteristic of pleurisy with effusion of tuberculous aetiology. Development of coarse adhesions interfers with resorption of the effusion (encapsulated pleurisy), while a prolonged purulent process may result in amyloidosis of the internal organs.

Resorption of effusion may be followed by some specific residual phenomena, such as sunken chest and the absence of diaphragmatic mibility on the affected side, displacement of the mediastinal organs toward the affected side, and sometimes permanent pleural friction.

Treatment

This, in the first instance, includes the therapy of the main disease, such as rheumatism (salicylates, amidopyrine, corticosteriods), pneumonia (sulpha drugs, and antibiotics), tuberculosis (PASA, phthivazide, streptomycin, canamycin, and others). Symptomatic therapy includes general strengthening (vitamins, etc.), desensitizing preparations, and high-calorie diet. Thermal procedures (compresses, diathermy) are useful to accelerate the resolution process. If pleural fluid is not resorbed during 2–3 weeks, evacuation of the effusion is necessary. Purulent exudate should obligatory be removed. Withdrawal of the fluid should be slow to avoid collapse or faint. As a rule, 0.5–1l of effusionis removed and antibiotics are injected instead into the pleural cavity. In order to acclerate resorption of the effusion, diuretics can be given. In the presence of cardiac failure, cordiamine, strophanthine and similar preparations are indicated. To prevent pleural adhesion during resorption of the exudate, remedial exercises should be prescribed.

Prophylaxis

Prevention of serofibrinous pleurisy consists in early diagnosis and active treatment of tuberculosis, rhematism and other diseases which may provoke pleurisy, and also in strengthening of the body (exercises, cool shower, etc.).

CHRONIC PNEUMONIA

Aetiology and pathogenesis

Chronic pneumonia often develops as a result of protracted bronchopneumonia. Its development is facilitated by frequent recurrent inflammations in the bronchi and the lungs, influenza epidemics, inhalation of harmful chemicals, smoking, etc. Involvement of autoimmune processes, helps conversion of acute and subacute diseases of the lungs into the chronic form. In most cases no specific causative agents are found in chronic pneumonia; streptococci, staphylococci, and pneumococci would be usually revealed by inoculation tests. If pneumonia is cured incompletely, the inflammatory process becomes slow (interstitial pneumonia) to give finally focal or diffuse pneumosclerosis. Bronchiectasis is formed in such patients,or they may develop emphysema and respiratory insufficiency.

Pathological anatomy

Areas of chronic inflammation in the lungs are observed along with the growth of connective tissue (focal or diffuse). Inflammatory changes may involve the bronchial walls and pleura; pulmonary vessels become stenosed and even obliterated.

Clinical picture

Many patients with chronic pneumonia (especially the agaed) may have normal temperature, the amount of expectorated sputum may be small, and there may be no changes in the blood. Dyspnoea, subfebrile temperature, permanent cough with expectoration of purulent or mucopurulent sputum, mild leucocytosis are found in some patients. Percussion may fail to reveal loss of resonance over the affected side due to a concurrent emphysema. Ausculation of the affected side reveals fine and moderate moist rales, sometimes againts the background or dry diffuse rales. X-ray examination is important: the interlobar septa are thickened, the lung pattern in the region of the inflammation is intensified, and the lung root is changed (enlarged lymph nodes).

Course

The course of the disease is usually protracted and slowly progressing. Three stages of the disease are distinguished in this country. The first stage of chronic pneumonia is protracted pnpeumonia (lsting over 6 weeks) and chronic bronchitis concurrent with relapsing pneumonia. The second stage is characterized by frequent exacerbations of the inflammation in the lungs alternating withmore or less prolonged remissions in the presence of symptoms of pneumosclerosis, lung emphysema and bronchiectasis. The third stage is characterized by marked symptoms, frequent exacerbations and pronounced functional disorders of external respiration and circulation.

Treatment

Combined treatement is necessary. Broad-spectrum antibiotics should be used with their frequent alternation to prevent microbial resistance to a particular antibiotic. For example, oletetrin is given per os for 5–10 days (250 mg 4–6 times a day), then it is replaced by the sodium salt of ampicillin (250–500 mg 4–6 times a day intramuscularly). Combination of antibiotics with sulpha drugs is effective. Bronchial draining is necessary in addition to the antibiotic therapy. It can be stimulated by giving expectorants, such as thermopsis preparations or ammonium chloride. In certain cases, corticosteroids are prescribed together with antibiotics to suppress auto-immune processes.

BRONCHIECTASIS

This is a conditions characterized by dilation of the bronchi. Bronchiectatic conditions are divided into primary (congenital, which are very rare) and secondary (secondary to various diseases of the bronchi, lungs and pleura).

As an independent (and frequently occurring) disease bronchiectasis has a specific clinical picture. It develops as a result of infection of bronchiectases and in the presence of chronic inflammation in them. Bronchiectasis may also be regarded as a form of chronic non-specific pneumonia. The disease occurs at any age, but more commonly between the ages of 20 and 40. The incidence among men is 6–7 times higher than it is in women.

Aetiology and Pathogenesis

Inflammation of bronchi and development of bronchiectasis in children can be the result of repeted acute bronchitis, whopping cough, measles, diphtheria, and sometimes tuberculous bronchoadenitis. Bronchiectases in youths and adults are formed due to acute diffuse bronchitis that develops against the background of chronic relapsing bronchitis, non resolved pneumonia, and also lung abscess, in recurrent pneumonia, and pulmonary tuberculosis. Bronchiectases develop in bronchitis only when the inflammatory process extends onto the muscular layer of the bronchial wall or onto all its layers. Mucle fibres are destroyed, the bronchus tone is lost at this area and its wall become thin. The absence of ciliated epithelium at the inflamed portions of the bronchus promotes accumulation of sputum in its lumen, upsets its draining function, and thus stimulates chronic inflammation. The inflamed site is first granulated but later connective tissue develops which disfigures the bronchus. Severely affected portions of the bronchi dilate during intense coughing.

Clinical picture

This depends on the size of bronchiectaes, their location, the degree of affection, and activity of the inflammatory process, the presence of emphysema of the lungs, and the degree of functional disturbances of external respiraton. If bronchiectasis affects the upper lobes of the lungs, the draining function of the bronchi remains intact, or is mildly affected. If the bronchiectases are located in the lower lobes of the lung, sputum is expectorated with difficulty to account for persistence of the inflammatory process.

The main symptom is cough with expectoration of seromucopurulent (three layers) or purulent sputum (sometimes foul-smelling). The daily amount of expectorated sputum varies from 50 to 500 ml and more. Blood streaks may be seen. Cough is paroxysmal in character and occurs mainly in the morning. The sputum accumulated during the night sleep irritates the sensitive nerve endings of the bronchial mucosa (due to alterations in posture). The sputum expectorated in the morning makes two thirds of the daily amount. Cough during the daytime is infrequent, and depends on accumulation of the sputum in the cronchiectases. Cough and expectoration of the sputum may also occur when the patient assumes the posture that stimulates the draining function of the ectatic bronchi. Haemoptysis, dyspnoea, excess sweating, weakness, headache, dyspepsia, deranged sleep and appetite, and wasting can also be observed. Bronchiectatic condition is exacerbated mostly in wet and cold weather: the body temperature rises, leucocytosis develops, and ESR increases. General inspection of the patient reveals acrocyanosis (at later stages of the disease) and oedematous face; in some cases the terminal phalanges of the fingers become clubbed (Hippocratic fingers), and the nails resemble the watch glass. The chest is of normal shape or emphysematous. Unilateral thoracid lagging occurs in the presence of unilateral bronchiectasis. Percussion sound is usually pulmonary, with a bandbox tone (due to a concurrent emphysema), less frequently with a tympanic tone (over the bronchiectatic area). The mobility of the lower border of the lung can be limited. Respiration is usually harsh or decreased vesicular (Due to emphysema). Dry and sometimes fine and moderate bubbling non-consonant rales are heard over the bronchiectatic ara. Pleural friction can be heard if inflammation extends onto the pleura and in the presence of pleural adhesions.

X-ray examination shows increased translucency of the lungs, deformation of the lung pattern, and the presence of bands in the lower lobes. Bronchography and tomography reveal the presence of bronchiectases and show their size and hsape.

Spirometry shows decreased vital lung capacity. In grave cases it decreases 2 or 3 times. The blood picture shows compensatory erythrocytosis and neutrophilic leucocytosis; the ESR may be increased, while at high erythrocytosis it may be decreased to 1–2 mm/h.

Course

The disease progresses if untreated. Anti-inflammatory therapy can cause a prolonged remission but the process may be exacerbated in a lapse of time (e.g. after overcooling).

Three stages of the disease are distinguished: the initial, moderately pronounced, and terminal. The final stage is characterized by marked changes in the internal organs; chronic right-ventricular heart failure (pulmonary heart), and amyloidosis of the liver, kidneys, and of other internal organs develop. Lung abscess, empyema of the pleura, pulmonary haemorrhage, and spontaneous pneumothorax may aggravate the disease.

Treatment

Broad-spectrum antiiotics are given intramuscularly, intracheally, and by inhalation. Antibiotic therapy can be combined with sulpha drugs. To improve the draining function of the bronchi, expectorants, broncholytics (ephedrine, theophedrine, euphylline), and also anti-allergic preparatons (in concurrent bronchospasms) should be given. The concommitant right-ventricular heart failure requires active cardiac therapy. Oxygen therapy and respiratory exercises are helpful.

If large saccular bronchiectases are localized only in one lobe, surgical treatment is indicated (reaction of the affected lobe).

Prophylaxis

Prevention of the onset and progress of bronchiectasis consists in regular medical check-ups of patients with chronic bronchitis and pulmonary fibrosis, their regular treatment, and control of harmful environmental and other effects (smoking, industrial hazards). Strengthening body of the body is also important.

EMPHYSEMA OF THE LUNGS

Lung emphysema is characterized by increased airiness of the lungs due to overdistended or destroyed alveoli.

Aetiology and Pathogenesis

The most common causes of emphysema of the lungs are obstructive bronchitis, chronic pneumonia, long-standing bronchial asthma, occupational lung diseases, etc. Lung emphysema occurs in mechanical overdistension of the lungs (in musicians playing wood-wind and brass instruments) or in heavy physical exertions associated with retention of breath. Advanced age is another predisposing factor.

Pathological anatomy

The pathological change in lung emphysema are characterized by destruction of interaveolar septa. The alveoli fuse to form bullae (bullous emphysema). The destroyed alveoli are not restored. The lungs are distended and lose elasticity.

Clinical picture

The patient mainly complains of dyspnoea, which at the onset of the disease may only develop during exercise but later it occurs at rest. Dyspnoea increases in cold seasons, in chills, and exacerbations of bronchitis; it is especially pronounced during attacks of cough. Dyspnoea is usually expiratory: a healthy person expires air whereas the patient with emphysema pressure it out from the chest with an effort. The intrathoracic pressure increases during expiration and the neck veins therefore become swollen. If heart failure concurs, the veins remain swollen during inspiration as well. Inspection reveals oedematous face, cyanotic mucous, checks, nose and the ear lobes; the skin is greyish. The terminal phalanges are often clubbed, and the nails lood like watch glass. In long-standing disease, the chest becomes barrel-shaped. Supraclavicular fossa are usually levelled or protrude over the clavicles. The tissue under the clavicles may protrude as well. Accessory muscles are actively involved in the respiratory act. During inspiration the rigid chest seems to rise due to contraction of the accessory muscles. A bandbox sound can be heard on percussion. Descending of the lower borders of the lungs and a limited mobility of the lower borders and the respiratory excursions of the lungs are characteristic of emphysema. Diminished vesicular respiration is heart on ausculation (the sounds are especially weakened in grave cases). In the presence of concurrent bronchitis, diffuse dry rales are heard.

The X-ray picture of the lungs is especially translucent. The lung borders are lowered, the mobility of the diaphragm is markedly limited.

The residual volume increases significantly in in emphysema while the maximum lung ventilation and the vital capacity decrease accordingly. The vital capacity may decrease 2.5–3 times and this intensifies the work of the respiratory apparatus (especially the respiratory muscles) and increases the oxygen demand. In order to decrease tissue hypoxia, compensatory mechanisms become actuated: cardiac rhythm is accelerated, the minute blood volume and erythrocyte counts are increased as well (compensatory erythrocytosis). Permanent intense cardiac activity in the lung emphysema with insufficient oxygen supply to the heart muscle is a cause of gradually progressing myocardial dystrophy which results finally in the right-ventricular failure. The respiratory insufficiency is then supplemented by heart failure with the specific clinical symptoms. The course of bronchitis and pneumonia in such patients is grave.

Course

Emphysema of the lungs usually progresses slowly. The patient may die from cardiopulmonary insufficiency.

Treatment and prophylaxis

These include treatment of bronchitis and pneumonia. Remedial exercise are especially important. Digoxin, ATP and other drugs stimulating the myocardium should be given in the presence of cardiac failure. Smoking control is also important in prophylaxis of emphysema.

2

Blood Circulatory System

MAJOR CLINICAL SYNDROMES

Cardiac Rhythm Disorders (Arrhythmias)

Any deviations from the normal rhythm of the heart are caled arrthythmias. These imply alterations in the heart rate, in succession or force of heart contractions, and also changes in the sequence of excitation and contraction of the atria and ventricles. Most arrhythmias are connected with functional changes or anatomical affections of the heart's conduction system.

The normal cardiac rhythm may change (1) in affected automaticity of the sino-atrial node, when the rate of sequence of impulses is altered; (2) in development of a focus of increased activity in the myocardium, that can generate impulses to initiate heart contractions apart from their generation in the sino-atrial node (ectopic arrhythmia); (3) in disordered conduction of the impulses from the atria to the ventricles or inside the ventricles themselves. Abnormal rhythm can also be due to impaired contractility of the myocardium. Arrhythmia can sometimes depend on changes in several functions of the heart such as automaticity, excitability, conduction or contractility.

Arrhythmias associated with altered automaticity of the sino-atrial node (sinus arrhythmia). When automaticity of the sino-atrial node is upset, the rate of impulse generation may either accelerate (sinus tachycardia) or slow down (sinus bradycardia), or the sequence of impulses may be changed with their genration at irregular intervals (sinus arrhythmia).

Sinus tachycardia is directly connected with effects of biologically active substances which increases excitability of the sino-atrial node. This phenomenon may also depend on the change in the tone of the vegetative nervous system. It develops with intensified effect of the sympathetic nervous sytem. The rate of cardiac contractions in sinus tachycardia usually varies from 90 to 120 and sometimes to 150–160 per min. Sinus tachycardia develops during meals, physical exertion and emotional stress. At elevated body temperature, the heart rate increases by 8–10 per min per each degree over 37°C. Sinus tachycardia is a frequent symptom of myocarditis, heart defects, and other diseases. It develops by reflex mechanism in heart failure and in response to the increased pressure in the orifices of venae cavae. Tachycardia often develops in neurosis, anaemia, hypotension, and in many infectious diseases and toxicosis; it can be provoked by some pharmacological preparations (adrenaline, caffeine, atropine sulphate, etc.), and in thyrotoxicosis.

The clinical signs of sinus tachycardia are heart palpitation and accelerated pulse. The T-P interval on ECG shortens and the P wave may interpose on the T wave.

Sinus bradycardia is connected with slowed excitation of the sino-atrial node, which in turn depends mostly on the increased influence of the parasympathetic nevous system on the heart (or decreased influence of the sympathetic nervous system). Automaticity of the sino-atrial node decreases in sclerotic affections of the myocardium and in the cold. The cardiac rate in sinus bradycardia decreases to 50–40 (in rare cases to 30) beats per min. Bradycardia may occur in well-trained atheletes. It is not permanent and the heart rhythm is acclerated during exercise as distinct from pathological bradycardia in atrioventricular block when bradycardia persists during and after exercise.If automaticity of the sino-atrial node sharply decreases (sick-sinus syndrome), the second-or third-order centres may function as the pacemaker, i.e. ectopic arrhythmias develop.

Sinus bradycardia occurs in increased intracranial pressure (tumour and oedema of the brain, meningitis, cerebral haemorrhage), in myxoedema, typhoid fever, jaundice, starvation, lead and nicotine poisoning, and due to effect of quinine and digitalis preparations. It may develop by reflex durign stimulation of baroreceptors of the carotid sinus and the aortic arch in essential hypertension, and can be provoked by pressure on the eye-ball (Dagnini-Aschner reflex), or by irritation of receptors of the peritoneum and the internal organs.

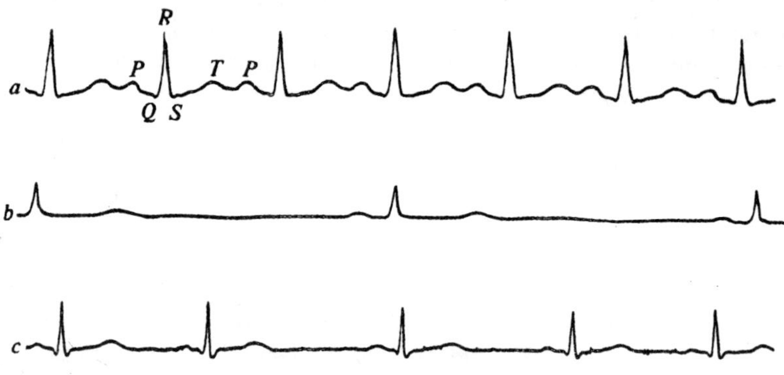

Fig. 2.1. Sinus arrhythmia.
a—sinus tachycardia (110 b.p.m.); b—pronounced sinus bradycardia (34 b.p.m.); c—sinus arrhythmia. cardiac complexes last for 0.70, 0.94, 0.82 and 0.86 s.

Mild bradycardia is not attended by any subjective disorders, nor does it produce any effect on the circulation. Marked bradycardia (under 40 beats per min) may cause nausea and loss of consciousness due to *cerebral anaemia*. Objective examination reveals slow pulse. The ECG in sinus bradycardia reveals the unchanged atrial or ventricular complexes; the T-P interval only increases to show protraction of electrical diastole of the heart; the P-Q interval sometimes increases insignificantly (to 0.20–0.21 s).

Sinus arrhythmia

Sinus arrhythmia characterized by irregular generation of impulses is due to variations in the tone of the vagus. It would commonly be associated with respiratory phases (respiratory arrhythmia): the cardiac rhythms accelertres during inspiration and slows down during expiration. Sinus arrhythmia is observed in children and adolescents (juvenile arrhythmia), in patients convalescing from infectious diseases, and in certain diseases of the central nervous system. It can be sign of pathology in rare cases when arrhythmia is not connected with respiration or when it develops in the aged during normal respiration.

Clinically sinus arrhythmia is not attended by any subjective disorders. The cardiac rhythm and pulse rate only change with respiratory phases, and the intervals between the heart complexes (R-R-R intervals) vary in length on the ECG.

Ectopic Arrhythmias

Additional (*heterotopic* or *ectopic*) foci of excitation can arise at any site of the conduction system (in the atria, ventricles, atrioventricular region). They can cause premature contraction of the heart before termination of the normal diastolic pause. This premature contraction is called extrasystole, and the disorder of the cardiac rhythm is called extrasystolic arrhythmia. If the activity of the of the ectopic focus is very high, it can become a temporary pacemaker, and all impulses governing the heart will during this time be emitted from this focus. The cardiac rhythm is then markedly acceleratred. The condition is known as paroxysmal tachycardia. Ectopic arrhythmias are often due to increased excitability of the myocardium.

The phenomenon known as re-entry can be another mechanism of ectopic arrhythmia. If an impulse meets an obstacle in the pathway of its conduction (local conduction disorder), the excitation wave can return from this obstacle to excite the myocardium.

Extrasystolic arrhythmia

Extrasystole usually develops during normal contraciosn of the heart governed by the sino-atrial node (nomotopic contractiosn). Ectopic foci of excitation can arise at any site of the conduction system. Usually excitatiosn arise in the ventricles, less frequently in the atria, the atrioventricular node, and inthe sino-atrial node (*sinus extrasystole*). A nomotopic contraction of the heart that follows extrasystole occurs in a longer (than normal) lapse of time. This can be explained as follows. During the atrial extrasystole, excitation from the ectopic focus is transmitted to the sino-atrial node to "discharge" it, as it were. The next impulse arises in the sino-atrial node only in a lapse of time that is required to "discharge" the node and to form a new impulse.

In ventricular extrasystole, the time between the extrasystolic contraction and subsequent nomotopic contraction is even longer. The impulse from the heterotopic focus, located in the ventricles, propagates only over the ventricular myocardium; it would not be usually propagated to the atria via *Aschoff-Tawara node*. The impulse occurs in normal time in the sino-atrial node but it is not transmitted to the ventricles because they are refractory after the extrasystolic excitation. The next impulse from the sino-atrial node will only excite and contract the atria and the ventricles. A long "*compensatory*" pause therefore follows the ventricular extrasystole which lasts till the next nomotopic contraction.

Extrasystolic arrhythmia is quite common. It may occur in practically healthy individuals as a result of overexcitation of certain sites of the conduction system due to the action of the extracardiac nervous system in heavy smokers adn in persons abusing strong tea or coffee; it can occur by reflex in diseases of the abdominal organs. Extrasystole often attends various cardiovascular pathological conditions due to inflammatory or dystrophic affections of the myocardium or its deficient blood supply; or it may be due to hormonal disorders (thyrotoxicosis, menopause), various intoxications, disorders of electrolyte metabolism, etc.

Patients with extrasystole can feel their heart missing a beat (escape beat) and a subsequent strong stroke. Auscultation of the heart reveals its premature contraction with a specific loud first sound (due to a small diastolic filling of the ventricles). Extrasystole can be easily revealed by feeling the pulse: a premature weaker pulse wave and a subsequent long pause are characteristic. If extrasystole follows immediately a regular contraction, the left ventricle may be filled with blood very poorly and the pressure inside it may be so small that the aortic valve would not open during the extrasystolic contraction and the blood will not be ejected into the aorta. The pulse wave on the radial artery will not be then detectable (missing pulse). The ECG of all extrasystoles are characterized by: (a) premature appearance of the cardiac complex; (2) elongated pause between the extrasystolic and subsequent normal contraction. According to the site of origin, extrasystoles are classified as atrial and atrioventricular (nodal), which are given a common name of supraventricular, and also ventricular (left- and right-ventricular) extrasystoles.

Excitation of the atria only changes in atrial extrasystole because the impulse is generated not in the sino-atrial node, and the ventricles are excited by the usual way. The ECG of atrial extrasystole is charactreized by the following signs: (1) Premature appearance of the cardiac complex; (2) preservation of the atrial P wave which may be slightly disfigured and superimposed on the preceding T wave; this depends on the abnormal atrial excitation form a heterotopic foucs; (3) normal shape of the ventricular complex; (4) slight elongation of the diastolic pause (T-P interval) following the extrasystolic contraction.

In atrioventricular (nodal) extrasystole the excitation of the atria differs from normal more substantially than in atrial extrasystole. The Aschoff-Tawara node impulse is transmitted to the atria retrogradely, from bottom to top. The ventricles are excited in nodal extrasystole in the usual way. The following signs are characteristic of the ECG in nodal extrasystole: (1) premature appearance of the cardiac complex; (2) change in the P wave which becomes negative to show the retrograde atrial excitation (in some cases the P wave is absent on the ECG); (3) the position of the P wave with respect to the ventricular complex changes, which depends on the rate of propagation of the excitation wave onto the atria and the ventricles. If excitation of the atria is followed by excitation of the ventricles, the negative P wave is recorded before the QRS complex; if the ventricles are excited first, the negative P wave follows the QRS complex. If the arteria and ventricles are excited synchronously, the P wave is not recorded separately but superimposes the QRS complex to alter somewhat its configuration. In other cases, the configuration of the ventricular complex in nodal extrasystole usually does not change; the diastolic pause becomes longer.

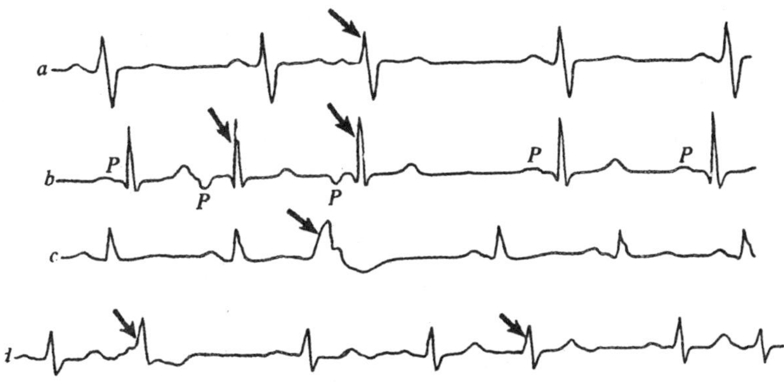

Fig. 2.2. Extrasystole.
a—atrial; b—nodal; c—ventricular; d—polytopic.
Extrasystoles are marked by the arrows.

Heart excitation order changes sharply in ventricular extrasystole. First the ventricular impulse is not usually transmitted retrogratedly through the ASchoff-Tawara node and the atria are not therefore excited. Second, the ventricles ar not excited synchronously (as in normal cases), but one after another, i.e. that ventricle is excited first where the ectopic focus is located. The time of excitation of the ventricles is therefore longer and the QRS complex wider. The ECG is characterized by the following signs: (1) premature appearance of the ventricular complex; (2) absence of the atrial P wave; (3) deformation of the QRS complex due to its increased voltage and length; (4) since the sequence of relaxation in the ventricles changes, the shape and the height of the

T wave changes as well. As a rule, the T wave is enlarged and its direction is opposite to that of the maximum wave of the QRS complex (the T wave is negative if the R wave is high, and positive if the S wave is deep). The ventricular extrasystole is followed by a long (full) compensatory pause (except in interpolated extrasystoles): the atria are only excited by the sinus impulse that follows the extrasystole because the ventricles are refractory at this moment. The P wave corresponding to the atrial excitation is "lost" in the disfigured extrasystolic ventricular complex. Only next (second to the extrasystole) sinus impulse excites both the atria and the ventricles, while the ECG shows a normal cardiac complex.

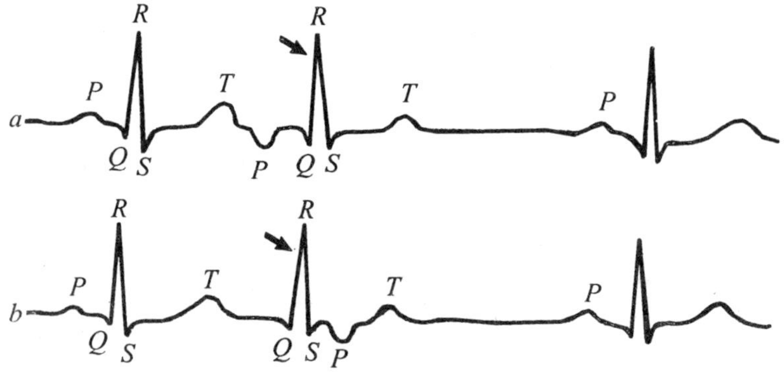

Fig. 2.3. Nodal Extrasytoles.
a—from the upper part of the sino-atrial node; ; b—from the lower poart of the node.

Some times it is possible to determine in which partiular ventricle the ectopic focus is located. This can be done from the configuration of the ventricular complex in various ECG leads. Left-venricular extrasystole is characterized by a high R wave in the third standard lead and the deep S wave in the first lead. In right-ventricular extrasystole, the extrasystolic complex is characterized by a high R wave in the first lead, and a deep S wave in the third lead.

Chest leads are very importnt for the topic diagnosis of ventricular extrasystole. Left-ventricular extrasystoles are characterized by the appearance of the extrasystolic complex with a high R wave in the right chest leads and a broad or deep S wave in the left chest leads. In righ-ventricular extrasystole, on the contrary, the deep S wave is recorded in the right chest leads, and a high S wave in the left chest leads. If excitability of the myocardium is high, several (rather than one) ectopic foci may exist. Extrasystoles generaed in various heart

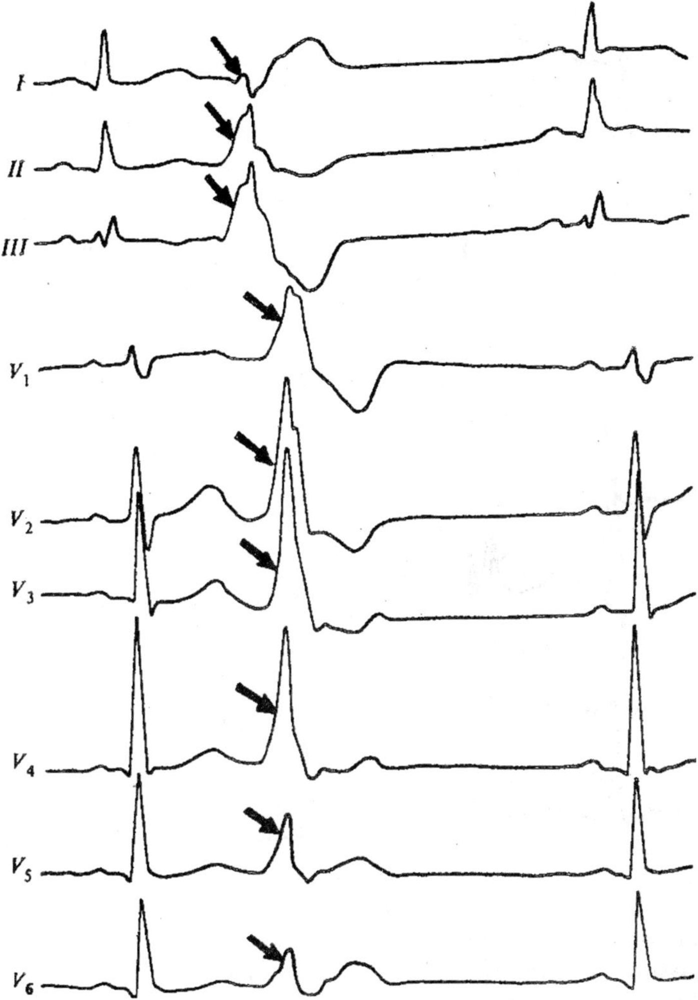

Fig. 2.4. Left-ventricular extrasystole.

chambers and having different configuration then appear on the ECG (*polytopic extrasystole*).

Wherever an ectopic focus may arise, its impulses may altrnate in a certain order with the normal impulses of the sino-atrial node. This phenomenon is known as *allorhythmia.* Extrasystole may alternate with each sinus impulse, or its may follow two normal impulses, or three normal impulses (quadrigeminy), etc. If the heterotopic focus is

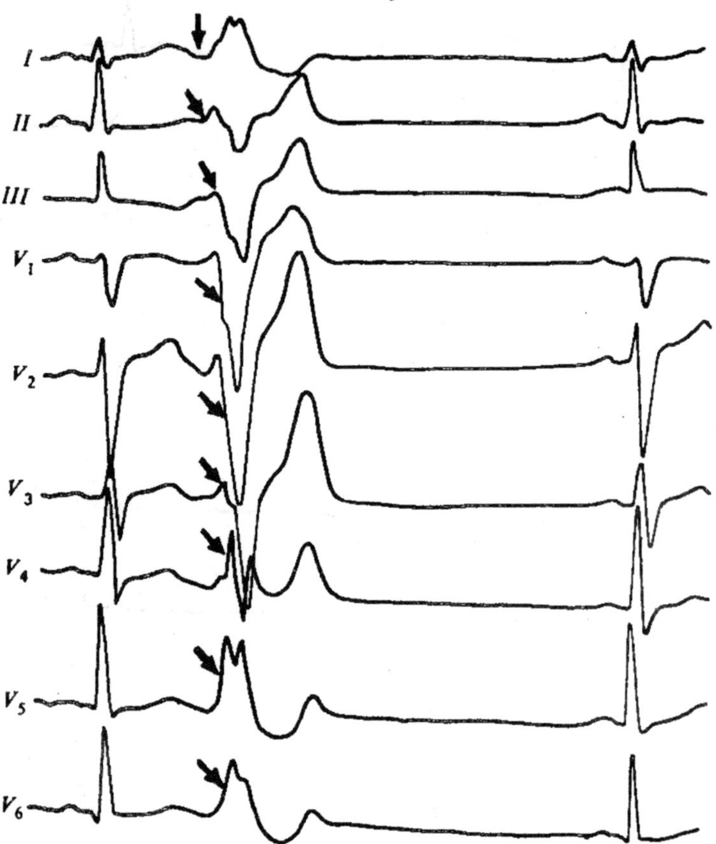

Fig. 2.5. Right-ventricular extrasystole.

even more active, a normal contraction may be followed by several extrasystoles at a run (group extrasystole) which sometimes precedes an attack of paroxysmal tachycardia.

Paraxysmal tachycardia

This is a sudden accleration of the cardiac rhythm (to 180–240 beats per min). At attack of paroxysmal tachycardia may last from several seconds to a few days and terminate just as unexpectedly as it begins. During an attack, all impulses arise from a heterotopic focus because its high activity inhibits the activity of the sino-atrial node. Paroxysmal tachycardia (like extrasystole) may occur in subjects with increased nervous excitability, in the absence of pronounced affections

Fig. 2.7. Ventricular trigeminy.

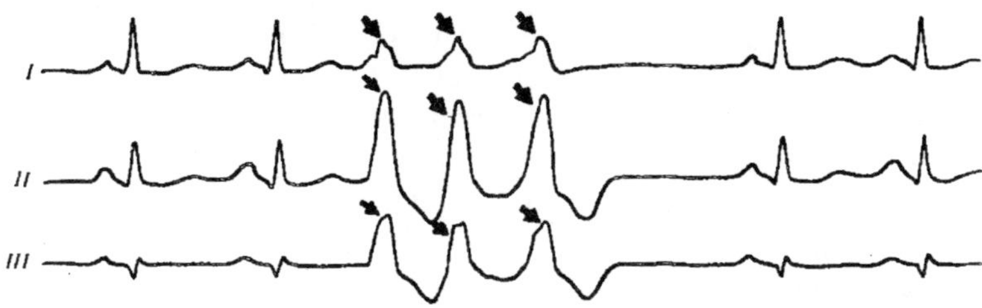

Fig. 2.8. Group extrasystole. Two normal cardiac complexes are followed by three ventricular extrasystoles.

of the heart muscle, but it arises more likely in the presence of a severe heart disease (e.g. myocardial infarction, heart defects or cardiosclerosis).

During an attack of paroxysmal tachycardia, the patient feels strong palpitation, discomfort in the chest,and weakness. The skin turns pale, and if attack persist, cyanosis develops. Paroxysmal tachycardia is characterized by swelling and pulsation of the neck veins, because during accelerted pulse (to 180–200 per min) the atria begin contractign before the ventricular systole ends. The blood is ejected back to the veins from the atria to cause pulsation of the jugular veins. Ausculation of the heart during an attack of paroxysmal tachycardia reveals decreased diastolic pause, whose length nears tht of the systolic one, and the heart rhythm becomes foetal (pendulum). The first sound increases due to insufficient ventricular diastolic filling. The pulse is

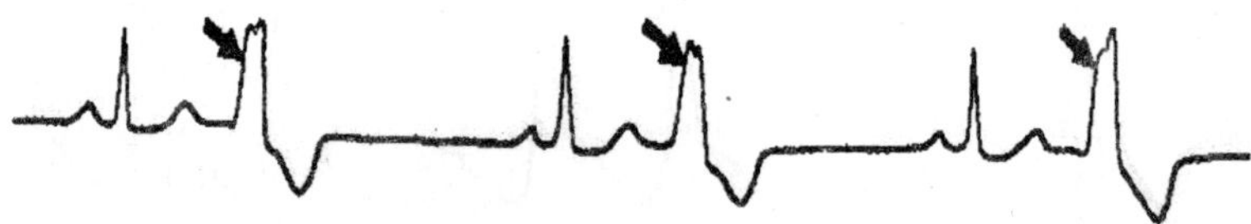

Fig. 2.6. Ventricular bigeminy.

rhythmic, very fast, and small. Arterial pressure may fall. If an attack persists (especially if it develops in the presence of a heart disease) symptoms of cardiac insufficiency develop.

Like in extrasystole, the heterotopic focus in paroxysmal tachycardia may be located in the atria, the atrioventricular node, and the ventricles. It is possible to locate the focus only by electrocardiography: a series of extrasystoles follow on an ECG at regular intervalsand at a very fast rate. Figure shows the ECG taken during an attack of supraventricular paroxysmal tachycardia (the P wave cannot be seen because of accelerated cardiac rhythm and the shape of the ventricular complex is not changed); ;an ECG which follows next is taken in a patient with ventricular tachycardia; it shows a series of altered and broadened ventricular complexes (similar to those in ventricular extrasystoles).

Arrhythmias dueto Disordered Myocardial Conduction

Transmission of the impulse may be blocked at any part of the heart's conduction system. The following types of heart blocks are distinguished: (1) sino-atrial block, in which beats are sometimes missing in the sino-atrial node and the impulse is not transmitted to the atria; (2) intra-atrial block, in which transmission of excitation through the atrial myocardium is impaired; (3) atrioventriclar block, in which

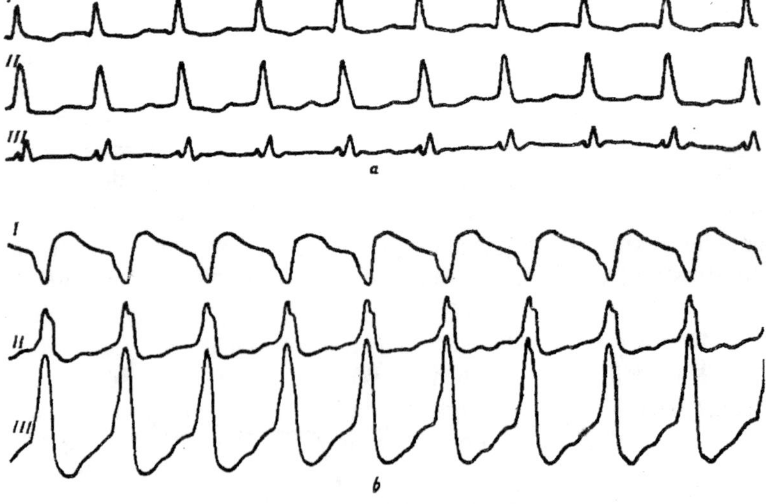

Fig. 2.9. Paroxysmal tachycardia.
a—supraventricular paroxysmal tachycardia (170 b.p.m.);
b—ventricular paroxysmal tachycardia (170 b.p.m.)

conduction of impulses from the atria to the ventricular is impaired; (4) intraventicular block, in which conduction of impulses through the His bundle and its branches is impaired.

Block may develop in inflammatory, dystrophic, and sclerotic affections of the myocardium (e.g. rhemuatic and diphtheritic myocarditis or cardiosclerosis). The conduction system may be affected by granulomas, cicatrices, toxins, etc. Conduction isoften impaired in disordered coronary circulation, especially in myocardial infarction (the interventricular septum is involved). Block may be persistent and intermittent. Persistent block is usually connected with anatomic changes in the conduction system, whereas intermittent block depends largely on the functional condition of the atrioventricular node and the His bundle and is often connected with increased influence of the parasympathetic nervous system; atropine sulphate is an effective means that restores conduction.

Clinical signs of block depend on its location. *Sino-atrial block* is characterized by periodic missing of the heart beat and pulse beat. The ECG shows periodic missing of the heart complex in the presence of a regular sinus rhythm (neither P wave nor the QRST complex are recorded); the length of diastole doubles.

Intra-atrial block can only be detected electrocardiographically because clinical signs are absent. Figure shows the ECG with altered P waves; since the time of atrial excitation inceases, the length of the P waves increases as well (to 0.1 s).

Attriventricular block is most important clinically. It is classified into three degrees gravity. The first degree an only be revealed electrocardiographically by the increased P-Q interval (to 0.3-0.4 s and more). block cannot be detected clinically, except that splitting of the first sound may sometimes be detected by auscultation (splitting of the atrial component). The second-degree atrioventricular block is characterized by dualismof its signs. Conductionof the Aschoff-Tawara node and His bundle is impaired: each impulse transmitted from the atriatot he ventricles increases and the P-Q interval on the ECG becomes longer with each successive beat. A moment arrives at which one impulse does not reach the ventricles and they don ot contract, hence the missing QRS complex on an ECG. During a long diastole, which now follows, the complex on an ECG. During a long diastole, which now follows the conduction power of the of the atrioventricular system is restored, and next impulses will again be transmitted, but their gradual slowing down will be noted again; the lengthof the P-Q

Fig. 2.10. Sino-atrial block. The thrid cardiac complex is followed by a pause equal to two preceding R-R intervals.

interval will again increase in each successive complex. The length of diastole which follows the P wave is called the Samoilov-Wenckebach period. This type of block is characterized clinically by periodically missing ventricular contractions, and hence missing pulse beats, which correspond to the Samoilov-Wenckebach period. On the other hand, the second-degree is trioventricular block can be characterized by a worse conduction. The P-Q interval remaisn constant, but only each second, third, or (less frequently) fourth impulse is transmitted to the ventricular complexe. This is knwon as incomplete heart block with a 2:1, 3:1, etc. ratio. Considerable decelerationof the ventricular rhythm and slow pulse are characteristic, especially in 2:1 block. If each third or foruth beat is missing,the pulse is irregular and resembles trigeminy or quadrigeminy with early extrasystoles and pulse deficit. If the heart rhythm slows down significantly, the patient may complain of giddiness, everything going black before his eyes, andtransient loss of consciousness due to anaemia of the brain. The third-degree atrioventricular block is called complete heart block. Atrial impulses do not reach the ventricles and the sino-atrial node becomes the only pacemaker for the aria. The ventricles contract by their own automaticity in the centresofthe second or third order. The number of their contractions in complete heart block is about 30–40 per min, and ventricular rhythm slows down with the lower position of the pacemaker in the conduction system.

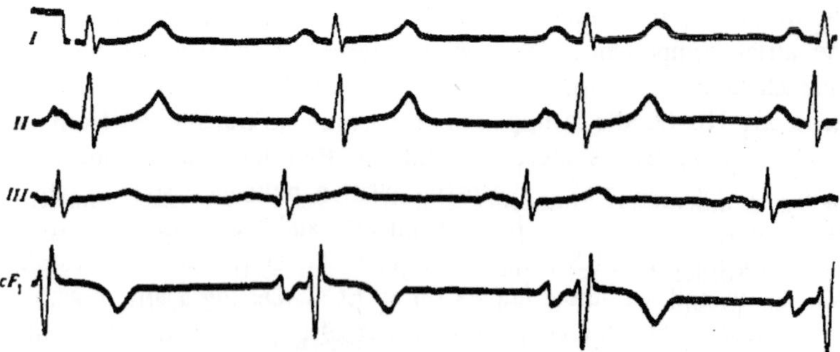

Fig. 2.11. Intra-artial block. P waves are broadened (P = 0.14 s) and serrated; the P wave in the first chest lead has two phases.

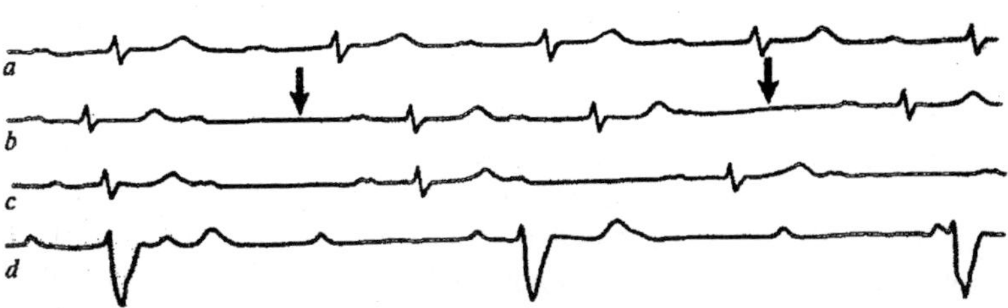

Fig. 2.12. Atrioventricualr block.
a—I degree (the P-Q interval in all cardiac complexes is 0.40 s); b—IIa degree with Samoilov-Wenckebach periods (marked by arrows); the P-Q interval in the first cardiac complex is 0.36 s, then follows the P wave; the ventricular complex is not recorded after the P wave; the next P-Q interval is 0.28 s and then 0.38 s; the next P wave is again followed by the Samoilov-Wenckebach period; c—IIb degree block with the ratio of 2:1; the artial rhythm is 84 c.p.m., and the ventricular rhythm is 42 c.p.m.; d—complete heart block; atrial rhythm is 85 c.p.m.; the ventricular rhythm 20 c.p.m.

The ECG in complete heart block is characterized by the following signs: (1) atrial P waves and ventricular complexes are recorded independently of each other, and part of the P waves may superimpose the QRS complex and become invisible on the ECG; (2) the number of ventricular complexes is usually much smaller than the number of atrial waves; (3) if the pacemaker arises from the Aschoff-Tawara node or His bundle, the shape of the ventricular complex does not change substantially; with lower location of the pacemaker in the conduction system, the QRST complexes are altered because the process of ventricular excitations is upset.

The heart rate in persistent complete heart block may be sufficiently high (40–50 beats/min) but the patient may be unaware of the disease for a long time. Examination of such patients reveals slow, thythmic, and full pulse. The heart sounds are dulled but a loud first sound ("pistol-shot" sound according to Strazhesko) may be heard periodically. It occurs due to coincidence of the atrial and ventricular contractions. If the ventricular rhythm slows down significantly (to 20 beats/min and less), or the heart misses a beat when incomplete heart block converts intoa complete one, i.e. when the impulses from the atria are not conducted to the ventricles, while their automaticity has not yet developed, attacks (the Morgagni-Adams-Stokes syndrome) may occur due to disordered blood supply, mainly to the central nervous system. During an attack the patient loses consciousness, faills, general epileptiform convulsions develop, the respiration becomes deep, the skin pallid, the pulse very slow or even impalpable. When the ventricular automticity restores, the patient regains his consciousness and all other signs of the syndrome disappear. If automaticity is not restored for a time, fatal outcome is possible.

Intraventricular block usually develops as the right or left bundle-branch block. The left limb of the His bundle ramifies almost immediately to give left anterior and left posterior branches. Only one branch can therefore be blocked. Block of the right limb may be combined with block of the branches of the left limb. In complete block of either of the limbs, the impulse from the sino-atrial node is normally conducted through the Aschoff-Tawara node and the main part of the His bundle to meet an obstacle to its conduction in that ventricle whose branch is affected. The ventricle with the intact branch is therefore first excited and excitation is transmitted tothe ventricle with the affected branch. The venticles are thus excited slowly and in anunusual way.

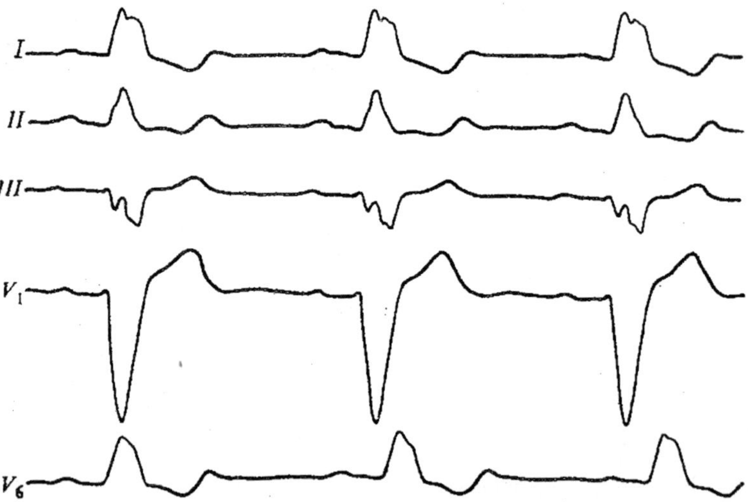

Fig. 2.13. Block of the left branch of the His bundle (time of intraventricular conduction is 0.17 s)

The bundle-branch block is characterized electrocardiographically as follows: (1) the P wave does not change; (2) the ventricles contract rhythmically by the impulse from the sino-atrial node, bu since the order of the ventricular excitationis upset, markedly altered and broadened QRS complexes (which resemble complexes in ventricular extrasystole) are recorded; (3) the tiem of intraventricular conduction increases to 0.12–0.18 s and more.The shape of the ventricular complexes depends on the particular bundle branch whichis blocked. If the left branch is blocked, its excitation is delayed and the ventricular complexes acquire the shape of the right-ventricular extrasystolic complexes, i.e. the QRS complex broadens and changes in its shape, the S-T interval shifts, and the direction of the T waves changes to the opposite with respect to the direction of the maximum wave of the QRS complex. If the right branch of the bundle is affected, the shape of the ventricular complexes resembles that of left-ventricular extrasystoles. Bundle-branch block cna only be detected electrocardiographically. It has no subjective signs.

Reduplication or splitting of the heart sounds can sometimes be auscultated. These are dueto asynchronous contractions of the ventricles.

Atrial and Ventricular Flutter and Fibrillation

Fibrillationis otherwise known as complete or absolute arrhythmia. It arises in cases with suddenly increased excitation of the myocardium and simultaneous conduction disorders. The sino-atrial node fails to function as the pacemaker and many ectopic excitation foci (to 600–800 per min) arise in the atrial mycoardium, which becomes only possible with a marked shortening of the refractory period. Since conduction of these impulses is difficult, each of them only excites and causs contraction of separate muscular fibres rather than the entire atrium. As a result, minor contractions develop in the atrium (atrial fibrillation) instead of adequate atrial systole. The mechanism of fibrillation is not fully understood. It is believed that permanent circulation of the circular excitation wave in the atrial can account for the development of this disorder. Only part of the impulses are transmitted to the ventricles through the Aschoff-Tawara node. Since conduction of atrial impulses is irregular, the ventricles contract at irregular intervals to cause complete arrhythmia of the pulse. Depending on the conductability of the Aschoff-Tawara node, three forms of atrial fibrillation are distinguished; tachyarrhythmic, in which ventricles contract at a rate from 120 to 160 per min, bradyarrhythmic, in which the heart rate does not exceed 60 per min, and normosystolic, in which the ventricles contract at a rate of 60–80 per min.

Fibrillation is characteristic of mitral heart disease (especially of mitral stenosis), coronary atherosclerosis, thyrotoxicosis, etc. Fibrillation may occur as a permanent symptom or in attacksof tachyarrhythmia. Clinically fibrillation (bradyarrhythmia) may cause no subjective symptoms. Tachyarrhythmia is usually characterized by palpitation. Examination of the heart reveals complete irregularity of the heart contractions. Variations in the length of diastole account for variations in ventricular filling and hence in the intensity of the heart sounds. The pulse is also arrhythmic, pulse waves vary in height (irregular pulse), and pulse deficit often develops in frequent heart contractions. The ECG of a patient with fibrillation shows the following changes: (1) the P wave disappears; (2) multiple small waves appear which are designated by the letter 'f'; (3) ventricular complexes follows at irregular, their shape does not change substantially.

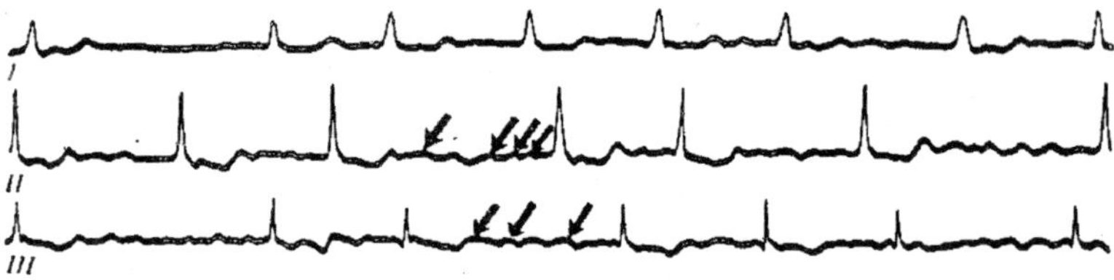

Fig. 2.14. Fibrillation. Ventricualr complexes follow on the ECG at irregular intervals; P waves are absent; small waves (f) are recorded instead.

Atrial flutter is the upset cardiac rhythm, which nears in its pathogenesis to fibrillation. As distinct from fibrillation, the number ofimpulses arising influtteringatria does not usually exceed 250-300 per min, and their conduction through the Aschoff-Tawara node is usually rhythmic. As a rule, not all atrial impulses are conducted to the ventricles. Each other, third or fourth impulse, is only conducted to the ventricles since partial (incomplete) atrioventricular block develops simultaneously. Conduction of the Aschoff-Tawara node sometimes constantly changes: each othe rimpulse is now conducted; then the rhythm changes to conduction of each third impulse, and the ventricles contract arrhythmically. Like fibrillation, atrial flutter occurs in mitral defects,k doronary atherosclerosis, and thyrotoxicosis; flutter sometimes develops in poisoning with quinine or digitalis.

Patients with accelerated heart rate (high conduction of the Aschoff-Tawara node) complain of palpitation. Examination reveals tachycardia that does not depend on the posture of the patient, exercise or psychic strain, since the sino-atrial node does not function as the pacemaker in atrial flutter (being governed by extracardial nerves). Heart contractions are arrhythmic in patients with varying conduction of the Aschoff-Tawara node. The ECG shows high waves instead of the normal atrial P waves. The number of high waves preceding each ventricular complex depends on the conduction of the Aschoff-Tawara node.

Ventricular fibrillation and flutter are gross disorders of the heart rhythm. The absence of adequae ventricular systole and contraction of separte ventricular muscles cause pronounced disorders in the haemodynamics and rapidly lead to death. Ventricular fibrillation and flutter occur in grave affections of the myocardium (diffuse myocardial infarction, etc.). The patient loses consciousness, becomes pallid, the pulse and arterial pressure become indeterminable. The ECG shows abnormal complexes on which separate waves are distinguished with difficulty.

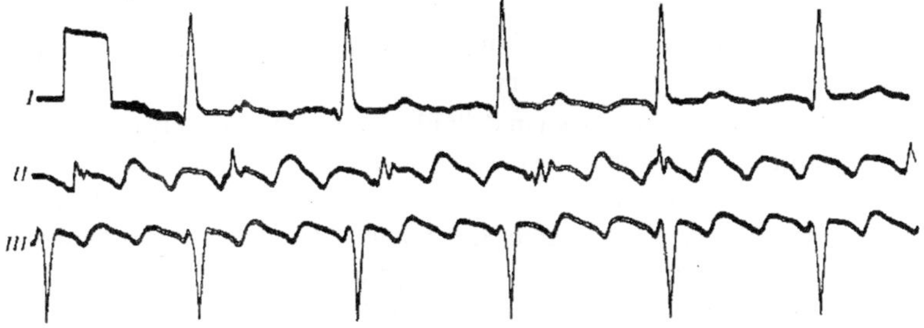

Fig. 2.15. Atrial flutter. High atrial waves are seen on the ECG.

Treatment of cardiac arrhythmia includes the following mesures: (1) management of the diseses which caused arrhythmia (mycarditis, ischaemic heart disease, neurosis, hyperthyroidism, etc.); (2) using means to restore ionic equilibrium in the myocardium and improve metabolism (potassium salts, vitamins, ATP, etc. etc.); (3) in cases with increased expreparations are recommended: quindine, novocainamide, aimaline, beta-adrenergic blocking agents (b-blockers), etc.; (4) progressive ventricular fibrillation is managed by electric defibrillation, i.e. short

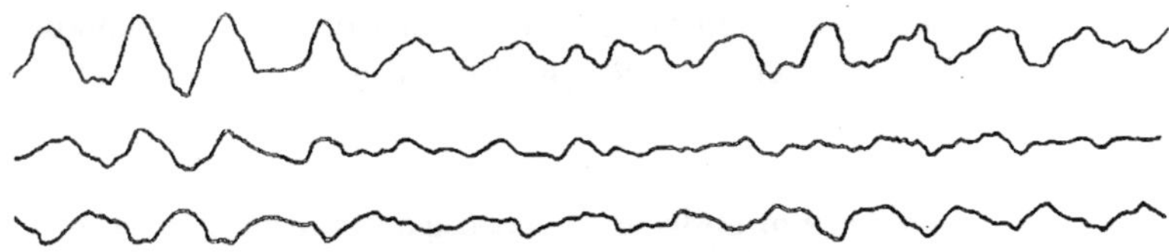

Fig. 2.16. Ventricular flutter and fibrillation.

(0.01 s) single electric discharge of 5000-7000 V, which causes instantaneous excitation of all parts of the myocardium adn restores the normal cardiac rhythm. Defibrillation is managed by an apparatus known as a deibrillator. Its two electrodes are attached to the chest (one below the left scapula and the other on the heart, or onebelow the right clavicle and the other over the heart apex). Electric impulses are also given to treat permanent fibrillation of paroxysmal tachycardia if medicamentous therapy proves inefficient; (5) electric stimulation of the heart (artificial packmaker) is also indicated in stoppage of the heart, in pronounced bradycardia, in complete atrioventricular block.

Circulatory Insufficiency

This is a pathological condition in which the cardiovascular system fails to supply the necessary amount of blood to the organs and tissues for their adequate function. This condition arises due to affection of the heart or of the vessels alone or it may be secondary to general disorders of the cardiovascular system. The clinic of circulatory insufficiency is usually associated with heart failure in which the function of the entire circulatory apparatus soon becomes affected.

Heart failure is associated with decreaed contactility of the myocardium. The amount of venous blood flowing to the heart and the resistance, which the myocardium has to overcome in order to eject blood into the vessels, exceeds the power of the heart to handlethe blood flowing from the veins into the atries. The numerous causes of heart failure can be classified into the following two major groups.

1. Heart failure caused by diseases which affect primarily the myocardium and its metabolism. This condition arises in

(a) infectious, inflammatory, and toxic affections of the myocardium (myocarditis of various aetiology, intoxication of the myocardium with alcohol, narcotic drugs, and other poisons);

(b) insufficient blood supply to the myocardium (disordered coronary circulation, anaemia);

(c) metabolic disorders avitaminosis, and endocrine dysfunction.

2. Heart failure due to overloadingor ovarstrain of the myocardium which arises in pathological changes in the heart proper or in the lood vessels (heart defects, hypertension in the greater or lesser circulation). The left or right ventricle, or the entire heart are overloaded due to various causes.

The left ventricle is overloaded

(1) in the presence of obstacles to blood ejection from the ventricle; this may be due to the narrowing of the aortic isthmus or orifice, or a sharp and persistent increases in the arterial pressure;

(2) in diastolic overfilling of the left ventricle in patients with aortic incompetence or mitral insufficiency.

The right ventricle may be overloaded (1) in difficult blood outflow from this ventricle due to the narrowing of the pulmonary trunk orifice, or increased pressure in the lesser circulation;

(2) in diastolic overfilling of the right ventricle associated with tricuspid or pulmonary valve incompetence.

Both ventricles are overloaded in combined or concurrent heart diseases, in some congenital heart diseases, in adhesive pericarditis, etc. Primary affections of the myocardium and its overloading sometimes contribute to the development of heart failure. Rheumatic myocarditis or rheumatic heart disease can, for example, become the cause of circulatory insufficiency in a rheumatic patient. heart failure may be aggravated by various infections and intoxications, physical strain, pregnancy, injuries, and surgical operations. The same factors can become a direct cause of heart failure in patients with heart disease, cardiosclerosis, etc.

Abnormal activity of the heart can for a long time be compensated by its intensified work and also by some extracardiac factors that ensure adaptation of the entire circulatory system to the increased demands of the body;

(1) the force of heart contractions is increased by the action of the vagus nerve;

(2) heart rate increases because elevated pressure in the orifice of the vena cava accelerates the cardiac rhythm;

(3) diastolic pressure decreases due to dilatation of arterioles and capillaries; this ensures a more effective systolic emptying of the heart;

(4) utilization of oxygen by tissue increases.

But prolonged overloading on the myocardium decreases the cardiac output and increases the residual systolic blood volume. This causes ventricular overfilling during diastole since the ventricle has to accept also its normal portion of blood.Diastolic pressure in the ventricle increases, the ventricle becomes distended, and the so-called tonogenic dilatation of the myocardium occurs. This dilatation, and the accompanying distention of the muscle fibres, intensify, according to Starling, contractility of the myocardium, cause its hyperfunction, and led to its hypertrophy. Compensatory hypertrophy of the myocardium increases the heart activity to maintain circulation for a long period of time.

Long-standing hypertrophy of the myocardium becomes the cause of its wear out, dystrophic, and sclerotic processes. These processes are also promoted by impaired blood supply to the myocardium, because hypertrophy of the heart is attended only by the increase in the weight of the myocardium, while the coronary vessels remain unchanged. The contractile power of the myocardium thus decreases and even its marked distention during diastole does not intensify contractility. Decrease in contractility and the tone of the myocardium are attended by a significant dilatation of the heart chambers, which is called *dmyogenic dilatation* (as distinct from the compensatory tonogenic dilatation). Myogenic dilatation can arise without preceding hypertrophy of the myocardium, in primary affections of the heart muscle (in myocarditis or myocardial infarction).

Tachycardia, which first develops as a compensatory mechanism to maintain the normal minute blood volume at decreased stroke volume, later becomes the cause of myocardial weakening, because diastole shortens in tachycardia and the time of restoration ofthe biochemical processes in the myocardium decreases.

Thus, dilatation and hypertrophy of the heart and tachycardia can only partially compensate for disorders in the cardiovascular system. As heart failure progress, these compensatory mechanisms themselves become the source of harmful effects. Further decrease in myocardial contractility causes marked disorders in haemodynamics.

Haemodynamic changes

As myocardial contractility decreases and myogenic dilatation of the ventricle develops, diastolic ventricular pressure increases while systolic pressure falls because the ability of the ventricle to strin during diastole sharply diminishes. This decreases the cardiac output and the minute blood volume. The mass of the circulating blood usually increases proportionally to the degree of circulatory insufficiency. This is favoured by retention of sodium chloride and water in decreased renal filtration and increased reabsorption of sodium, and the increasing number of red blood cells (hypoxia is attended by intensified haemopoiesis to compensate for the developing insufficiency). The rate of blood flow decreases. ***Blood pressure*** changes as well: venous and capillary pressure increases in the greater circulation; arte;rial pressure remaisn normal or diastolic pressure slightly increases and the pulse pressure decreases.

Haemodynamic disorders are attended by abnormal gas exchange. Slowing of the blood flow rate rate increases oxygen absorption by the tissues; to 60-70 per cent of oxygen (instead of the normal 30 per cent) is consumed in capillaries. The arteriovenous difference in the oxygen content of the peripheral blood increases. Developing disorders in gas exchange associated with heart failure upset the carbohydrate metabolism. Lactic acid that is formed in skeletal muscles is decomposed only partly in insufficient oxygen supply to tissues; the blood content of lactic and pyruvic acids therefore increases. The increased content of lactic acid upsets acid-base equilibrium and decreases alkaline reserve. At the beginning of heart failure acidosis is compensated for because lactic acid displaces carbon dioxide which is removed from the body through the lungs. If pulmonary ventilation is upset and carbon dioxide is not expired in sufficient amount, decompensated acidosis develops.

Accumulation of underoxidized metabolites in the blood and intensified work of respirtory muscles intensify basal metabolism to complete a vicious circle; the body's demands in oxygen increase, whereas the circulatory system fails to meet them. Oxygen debt thus increases.

Haemodynamic and metabolic disorders account for various clinical signs of heart failure.

Clinical manifestations of heart failure

An early and specific symptom of circulatory insufficiency is dyspnoea, It is manifested by unreasonably accelerated and intensified respiration; dyspnoea develops at rest or during mild exercise. Dyspnoea arises in upset gas exchange and accumulation of underoxidized metabolities in the blood. It has already been said that lactic acid accumulated in the blood combines with bicarbonate alkalis to displace carbon dioxide which stimulates the respiratory centre to accelerate and deepen respiration. Especially pronounced disorders in gas exchange arise in blood congestion in the lesser circulation when the respiratory surface and the respiratory excursions of the lungs decrease.

Development of dyspnoea is also provoked by accumulation of liquid in the pleural andthe abdominal cavities which interferes with the respiratory excursions of the lungs. During exercise dyspnoea markedly increases; it also becoles more pronounced after meals, and in the recumbent position of the patient. Grave dyspnoea sometimes occurs in attacks, which are called cardiac asthma.

Heart failure is often attended by cyanosis. The skin and the mucosa turn blue with increasing content of the reduced haemoglobin in capillary blood (over 50 g/l, or 5 g/100 ml), which, in contrastto oxyhaemoglobin, is dark. The dark blood is seen through the skin to colour it bluish; the colour is especially intense at sites where the skin is thinner (lips, cheeks, ear auricles). Cyanosis in circulatory insufficiency may be due to overfilling of vessels of the lesser circulation with blood and impaired arterialization of blood (the so-called central cyanosis). Peripheral cyanosis, however, occurs more frequently. It is connected with the slowing down of the blood flow and increased oxygen utilization by the tissues. Since the slowing down of the blood flow is more pronounced in parts of the body remote from the heart, the blue colour appears in the limbs, the ears, and the tip of the nose (acrocyanosis). Cyanosis is favoured by the widening of the venous network in the skin, increased volume of the circulating blood, and its increased haemoglobin content.

Oedema is an important sign of circulatory insufficiency. Its development in individuals with heart diseases depends on the following factors:

(1) increased hydrostatic pressure inthe capillaries and slowed blood flow which promote transduation of fluids into tissues;

(2) abnormal hormonal regulationof the water-salt metabolism. Insufficient supply of arterial blood to the kidneys intensifies excretion of renin which increases secretion of adrenal cortex hormone, aldosterone. Aldosterone in turn increases the reabsorption of sodium in convoluted renal tubules to promote retention of fluid in the tissues. Furthermore, secretion of antidiuretic pituitary hormone increases to intensify re-absorption of water. These disorders in water-salt metabolism increase the volume of blood plasma, venous and capillarypressure, and intensify transduation of fluid in tissues;

(3) during long-standing venous congestion in the great circulation, liver function decreases and the production of albumins becomes disordered to decrease oncotic pressure of blood plasma. Moreover, liver dysfunction inhibits the decomposition of the antidiuretic hormone and aldosterone in the liver.

Cardiac oedema can first be latent. Retention of fluid in the body (sometimes to 51 and more) does not immediately cause visible oedema but provokes a rapid gain in the patient's weight and his decreased urination. Oedema becomes visible in the first instance in the lower part of the body: in the loer limbs (if the patient sits or stands) and in the sacral region (if the patient keeps bed). If circulatory insufficiency is progressive, oedema increases and dropsy of the body cavities develops. Fluid can be accumulated in the abdominal cavity (ascities), in the pleural cavity (hydrothorax), and in the pericardial cavity (hydropericardium). Ascites may provoked by a prolonged venous congestion in the liver attended by its fibrosis and portal hypertension. Ascites then prevails over dropsy of other cavities.

Practically all other organs are changed in patients with heart failure. Changes in thelungs are connected with prolonged congestion of blood in the lesser circulation. Congested lungs become rigid to devrease respiratory excursions of the chest and limit mobility of the lower border of the lungs. The so-called congestive bronchitis thus develops. Patients develop cough; it may be dry or with expectoration of small amountsof mucous sputum; harshrespiration is heard over the lungs during ausculation; dry rales (intense in the postero-inferior parts of the chest) are heard; later they become moist. Long-standing venous congestion in the lesser circulation stimulates development of connective tissue in the lungs, which in turn impaires gas exchange. The overfillingof smaller vessels with blood may be accompanied by their rupture and the appearance of blood in the sputum. Insiginficant haemorrhage, and also insinuation of erythrocytes through the blood vessels, promote depositionof the blood pigment in the lungs and development of brown induration; heart-failure cells can be detected in the sputum.

There are also some cardiovascular signs indicating inadequate contractility of the myocardium. Considerable myogenic dilatation of the heart chambers can cause relative insufficiency of atrioventricular valves. The heart borders broaden, the sound of the heart (especially the first sound) weaken, tachycardia develops, and gallop rhythm sometimes occurs to suggest a grave affection of the myocardium and lowering of its tone. Organic murmurs usually weaken, because the blood-flow rate slows down; functional murmurs associated with relative atrioventricular incompetencemay develop.

The liver quickly responds to venous congestion in the greater circulation. It becomes enlarged, Glisson's capsule distended, and the patient complains of the right hypochondrium pain. If congestion develops gradually, the patient feels heaviness in the eipgastric and right hypochondriu. Prolonged venous congestion in the liver stimulates development of connective tissue (cardiac fibrosis of the liver) with subsequent hepatic dysfunction and portal hypertension.

The gastro-intestinal function is also impaired. Congestion in the greater circulation provokes congestive gastritis and intestinal dysfunction. Patients complain of poor appetite, nausea, and vomitting, they develop metorism and constipation. Dyspeptic and metabolic disorders cause disturbances in nutrition; as circulatory insufficiency further progresses, grave asthenia (cardiac cachexia) develops.

Venous congestion in the kidneys deceases the daily amount of the urine, and its specific gravity increases. A small amount of protein, red blood cells, and casts can be found in the urine.

Circulatory insufficiency soon becomes attended by dysfunction of the central nervous system. Rapid fatigue, decreased work capacity and mental power, high irritability, deranged sleep, and sometimes, depression are characteristic.

The clinical signs and changes in various organs of the body depend on the degree and duration of heart failure, and on the particular side of the heart that is affected (right or left).

Bearing these factors in mind,Strazhesko and Vasilenko provided their *classification of circulatory insufficiency*, which was adopted at the 12th All-Union Congress of Therapeutists in 1935. According to this classification, the following forms of circulatory insufficiency are distinguished.

1. Acute circulatory insufficiency. It can depend on acute heart failure (either side) or failure of any of its chambers (left or right ventricle, left atrium), or else it may be caused by acute vascular insufficiency (collapse or shock).

2. Chronic circulatory insufficiency. This can be divided into three stages.

The first stage (initial) is latent circulatory insufficiency, which is only manifested during physical exercixe, while at rest the haemodynamics and functions of the organs are normal; the work capacity is decreased.

The second stage is characterized by a pronounced prolonged circulatory insufficiency, haemodynamic disorders (congestion in the lesser or greater circulation) and sysfunction of organs at rest; the work capacity of patients is markedly decreased. Two periods are distinguished at this stage: (1) the initial period, with mild haemodynamic disorders; and (2) the final period characterized by grave haemodynamic disorders.

The third stage is the terminal or dystrophic stage of circulatory insufficiency. In addition to grave haemodynamic disorders, irreversible morphological changes develop in the organs along with persistent metabolic disorders and disability.

Clinical Forms of Heart Failure

Acute heart failure may develop in gravedisorders in the cardiac rhythm (paroxysmal tachycardia, ventricular fibrillation, myocardial infarction, acute myocarditis, and the like). Acute heart failure is attended by marked drop in the minute volume and filling of the arterial system; clinically it is very much like circulatory insufficiency of the vascular genesis (it is sometimes termed as acute cardiac collapse).

Clinically it is manifested by sudden and pronounced asthenia, sometimes by syncopes due to brain ischaemia, pallidness and cyanosis of the skin, cold limbs, small or thready pulse, and decreased arterial pressure.

The cardiac aetiology of circulatory insufficiency is confirmed by change in the heart proper (valve incompetence or arrhythmia, roadening of the heart borders, changes in the heart sounds and gallop rhythm). The attending venous congestion is manifested by dyspnoea, swelling of the neck veins, rales in the lungs, and enlargement of the liver. Acute heart failure may depend not on the weakening of the entire myocardium but on a pronounced decrease in contractile capacity of the myocardium of one of the heart chambers: left ventricle, left atrium, or right ventricle.

The syndrome of acute left-ventricular heart failure arises in patients in whom the left ventricle is mostly affected (essential hypertension, aortic incompetence, myocardial infarction). A typical symptom of acute left-ventricular heart failure is cardiac asthma (attacks of severe dyspnoea due to acute congestion in the lungs and upset gas exchange). Attacks can be provoked by physical exercise and nervous strain. Attacks usually occur during night sleep. This can be explained by an inceased vagus tone during sleep, which causes narrowing of the coronary arteries and thus impairs nutrition of the myocardium. Moreover, blood supply to the respiratory centre decreases during sleeep and its excitability diminishes. The lesser circulation becomes overfilled with blood because during a sharply decreased contractility of the left-ventricular myocardium, the right ventricle continues working intensely to pump the blood from the greater circulation to the lesser one.

During the attack of cardiac asthma, the patient develops asphyxia and marked weakness; cold sweat appears. He has to assuem a forced position—sitting with his legs hanging down from the bed (or he stands up). The patient begisn coughing and expectorates tenacious sputum. The skin becomes pallid and cyanotic. Moist and dry rales are heard over the lungs. The heart sound are weakened at the apex and over the pulmonary artery intensified (the second sound). Tachycardia and small frequent pulse are characteristic. If congestion in the lesser circulation progresses, the blood plasma and blood corpuscles pass from the overfilled pulmonary capillaries to the alveoli and accumulate in the respiratory ducts; oedema of the lungs develops to intensify still more or feeling of suffocation and cough; respiration becomes rattling; ample foaming sputum with traces of blood (pink or red) is expectorated.

Many moist rales of various calibres are heard over the lungs (over their entire surface). Ausculation of the heart often reveals gallop rhythm. Pulse is markedly accelerated and thready. Oedema of the lungs requires prompt and energetic measures to be taken to prevent possible death.

The syndrone of acute left-atrial failure develops in patients with mitral stenosis in markedly weakened contractility of the left atrium and normal function of the right ventricle, which continues pumping blood into the lesser circulation. This causes overfilling of its vessels with venous blood and development of the same clinical symptoms as those in acute left-ventricular failure.

The syndrome of acute right-ventricular failure is especially pronounced in embolismof the trunk of the pulmonary artery or its branches into which the thrombus is carried from veins of the greater circulation or from the right chambers of the heart. Patients breathe rapidly, cyanosis and cold sweat appear; they feel pressure and pain in the heart. The pulse becomes small and frequent, the arterial pressure drops. Acute right-ventricular failure causes a pronounced venous congestion in the greater circulation. The venous pressure increases; the neck veins become swollen and the liver enlarges; oedema laer develops.

Like acute failure,*chronic heart failure* may first depend mostly on failure of one of the heart chambers. The syndrome of chronic left-ventricular failure develops in many diseases attended by affections of the left ventricle (aortic incompetence, mitral failure, arterial hypertension, coronary insufficiency due to dystrophy of the left-ventricular muscle, etc.). The syndrome is characterized by a persistent blood congestion in the blood flow through the vessels of the lesser circulation is slowed, the gas exchange is upset, and dyspnoea, cyanosis, and congestive bronchitis develop. Blood congestion in the lesser circulation is even more pronounced in chronic left-atrial failure in patients with mitral stenosis. It is manifested by dyspnoea, cyanosis, cough, and haemoptysis. Prolonged venous congestion in the lesser circulation stimulates growth of connective tissue in the lungs and sclerosis of the vessels. Another, pulmonary barrier is thus produced to become an obstacle to normal passage of blood through the vessels of the lesser circulation. Pressure in the pulmonary artery elevates to increase the load on the righ ventricle, which later becomes the cause of the failure.

The syndrome of chronic right ventricular failure arises in mitral heart diseases, lung emphysema, pneumosclerosis, tricuspid incompetence, and in certain congenital heart defects. It is characterized by a marked venous congestion in the greater circulation. The patient is cyanotic, the skin sometimes becomes icterocyanotic. The peripheral veins, especially the neck veins, become swollen, the venous pressure increases, oedema and ascites develop, and the liver is enlarged.

Primary dysfunction of one of the heart chambers may eventually cause total heart failure, which is characterized by venous congestion in both the greater and lesser circulation. Moreover, chronic heart failure attended by dysfunction of the entire circulatory system arises in diseases affecting the myocardium (myocarditis, intoxication, ischaemic heart disease, etc.).

The severity of clinical signs in heart failure depends on the stage of circulatory disorders.

In the initial, latent stage of circulatory insufficiency, the patient's work capacity decreaes, physical exertion provokes dyspnoea, palpitation, and oxygen debt increases to a greater degree than that in healthy subjects. These symptoms subside at rest.

The second stage of circulatory insufficiency is characterized by haemodynamic disorders. In the initial period (stage A), the patient develops dyspnoea during normal exercise (e.g. in walking), and his work capacity deceases markedly. Examination reveals moderate cyanosis and pastous shins. Congestion in the lungs is not pronounced: the respiratory movements of the chest and excursions of the lower lung borders are decreased; the vital capacity of the lungs is diminished. The liver is mildly enlarged. The venous pressure increases. Stage B is characterized by marked congestion in the greater and lesser circulation. Dyspnoea develops even at rest which is intensified durign slight physical exertion. Patients are fully disabled. Typical signs of heart failure (pronounced cyanosis, oedema, ascites, dysfunction of various organs) are revealed.

The third stage is charcterized by pronounced metabolic disorders caused by a prolonged circulatory insufficiency. The patient would be extremely asthenic, with irreversible morphological changes in the lungs, liver, and kidneys. The combination of metabolic disorders in circulatory insufficiency was called by Vasilenko "circulatory dystrophy".

In additon of circulatory insufficiency, which ismanifested by haemodynamic disorders, some authors distinguish also energodynamic insufficiency (according to Hegglin). This depends on metabolic disorders in the myocardium and occurs in pronounced hypokaliaemia, diabetic coma, severe diarrhoea, toxicosis and infections, and is characterized by a considerable shortening of the mechanical systole of the heart. This can be revealed by the second sound which appears on a PCG (at the end of the T wave) earlier than in normal cases. The *Q-T* interval on the ECG (representing the electric systole of the ventricles) is on the contrary prolonged.

Vascular Insufficieny

Circulatory insufficiency of the vascular aetiology arises in cases where the equilibrium between the capacity of the vessels and the volume of the circulating blood is upset. It develops when the volume of blood is diminished (loss of blood, dehydration of the body), or when the vascular tone drops. Diminished vascular tone depends mostly on (1) reflex disorders in thevasomotor innervation of the injured vessels, irritation of serous membranes, myocardial infarction, embolism of the pulmonary arery, etc. (2) disordered vasomotor innervation of cerebral aetiology (hypercapnia, acute hypoxia of the diencephalon, psychogenic reactions); (3) vascular paresis of toxic origin which occurs in many infections and toxicosis. Diminished vascular tone disturbs normal distribution of blood in the body: the amount of deposited blood increases, especially in the vessels of the abdominal organ, whereas the volume of circulating blood decreases; the decreased volume of circulating blood wakens venous blood flow to the heart; the stroke volume of blood decreases and arterial and venous pressure diminish as well. Circulatory insuffiency of vascular genesis is usualy acute and is then called circulatory collapse. Decreased volume of circulating blood and reduced arterial pressure cause ischaemia of the brain; acute vascular insufficiency is therefore characterized by giddiness, darkening in the eyes, noise in the ears, and often by loos of consciousness (syncope). Objective examination of the patient reveals pallid skin, cold sweat, colb limbs, accelerated and superficial respiration, small and sometimes thredy pulse, and decreased arterial pressure.

A syncope is a symptom of acute vascular insufficiency. This is an abrupt and transient loss of consciousness dueto insufficient blood supply to the brain. A syncope may occur in fatigue, excitation, freight or stay in a non-ventillated room, etc. This phenomeno is connected with upset central nervous regulation of the vascular tone, as a result of which blood is accumulated in the vessels of the abdominal cavity. A patient in a syncope is covered with cold sweat, has a pallid skin,cold limbs, andsmall or thready pulse. Some patients lose consciousness when they suddenly change from lying to the upright position. It isespecially characteristic of young asthenic subjects, mostly women. Fatigue, anaemia, and infectious diseases also predispose to a syncope. Such a syncope is called orthostatic collapse. It is explined by a delayed response of the vasomotor apparatus, owing towhich blood flows from the upper portion of the body to the vessels of the lower limbs and the abdomen when an individual changes his posture.

Treatment of Patients with Circulatory Insufficiency

Causes of cardiovascular insufficiency should first to be removed. These may be rheumatism, essential hypertension, myocarditis, coronary insufficiency, or other diseases. If a heart disease, adhesive pericarditis, or hert aneurysm are the cause of circulatory insufficiency, possibility of surgical treatment should first of all be considered in order to remove or lessen the mechanical obstacles to normal work of the heart.

The patient should be given physical and mental rest. If heart failure is severe, the patient must keep bed. As the condition improves,dosed exercises are recommended, and remedial exercises are prescribed. The intake ofliquid should be limited to 500-600 ml a day. The amount of salt should also be limited to 1-2 g a day. The diet should be sufficiently rich in vitamins; it should not provoke flatulence.

The initial stages of circularoty disorders should be treated by rest and sedative preparations, which will restore normal function of the cardiovascular system. If circulatory insufficiency is significant, cardiac glycosides should be given; digitalis preparations (e.g. digoxin 0.00025 g twice a day, in tablets, per os); strophathin (0.5 ml of a 0.05 per cent solution, intravenously, slowly, in 20 ml of a 20 per cent glucose solution), and other preparations by the influence of which cardiac activity will be improved to increase the minute volume, to slow down the cardiac rhythm, and to decrease venous pressure. Diuretics are very important in the therapy of circulatory insufficiency. Most modern diuretics are saluretic, i.e. substances that decrease reabsorption of sodium and water in the renal tubules. Furocemide is often given (0.04 g, one tablet per os, or a solugion containing 0.02 mg of the active principle in 2 ml of a 1 per cent ampouled solution, intramuscularly). Other diuretics are also given. If venous congestion is significant, phlebotomy is indicated (200--400 ml). If fluid is accumulated in the pleural and abdominal cavities, it should be removed by puncture. Oxygen therapy is indicated to patients with circulatory disorders to decrease tissue hypxia. In order to improve metabolism in the myocardium, vitamins of group B, ascorbic acid, ATP, inosine, and potassium preparations are prescribed.

A patient with an acute vascular insufficiency should be placed in the horizontal position with the legs slightly elevated on a special collapsible bed. The cause of acute vascular insufficiency should be removed; haemorrhage should be arrested, anaesthetics given in shock, and the patient should be warmed. An isotonic sodium chloride solution, antishock fluid, blood substitutes (polyglucin, blood plasma, etc.) should be given in travenously. Blood transfusion is indicated in cases with loss of blood. The vascular tone is raised by caffeine, cordiamine, mesaton, and corticosteroid hormones.

SPECIAL PATHOLOGY

Diseases of the cardiovascular system stand among the first in the list of diseases of internal organs. They are a frequent cause of early disability and untimely death of patients. Most common are ischaemic heart disease, rheumatism and rheumatic heart diseases, endocarditis, myocarditis of various genesis, and essential hypertension.

Rheumatism

Rheumatism is a general infectious and allergic disease in which connective tissues, mainly of the cardiovascular system, are affected by inflammation; joints, serous membranes, internal organs, and the central nervous system are often involved. Rheumatism is a collagenous disease, i.e. a disease characterized by a systemic and progressive derangement of connective tissue.

Rheumatism was classified as an independent disease with typical affections not only of the joints but mainly of the heart in 1835 by a French clinicist Bouillaud and in 1836 by the Russian physician Sokolsky. Until that time rheumatism had been considered a disease of joints.

Aetiology and pathogenesis

Beta-haemolytic streptococus of group A is believed to be the causative agent of rheumatism. This conjecture is confirmed by (1) frequent incidence of rheumatism following streptococcal infection; (2) increased antibody titres to various antigens and enzymes of the streptococcus in the blood of rheumatic patients; (3) successful prophylaxis of rheumatism by antibacterial preparations.

The pathogenesis of rheumatism is complicated and not well studied. At the present time, the development of the disease is described as follows. Most person affected by streptococcus develop stable immunity. This immunity does not develop in 2-3 per cent of the affected subjects due weakness of their defence mechanisms and they become sensitized by the streptococcus antigen. In these conditions the infection re-enters the body to cause a hypergic response in connective tissues; clinical signs of the disease thus develop. Autoimmune processes are very important in the onset of rheumatism. The affected connective tissue acquire antigenic properties; auto-antigens (secondary antigens) cause formation of aggressive auto-antibodies. They affect not only the connective tissue that has already been affected by the primary antigen but also intact tissue to aggravate the pathology. Re-infection, cooling, and overstrain promote formation of new auto-antigens and auto-antibodies to strengthen the pathological reaction of the upset immunity and to provide conditions for the recurring progressive course of the disease.

Pathological anatomy

Four phages of derangement of connective tissue are differentiated in rheumatism: (1) mucoid swelling; (2) fibrinoid changes; (3) granulomatosis; and (4) sclerosis.

Mucoid swelling is characterized by supreficial derangement of connective tissue which involves mainly the interstitial substances and only insignificantly the collagenous complex. The process occurring at this stage are reversible. Derangement of connective tissue occurring at the fibrinoid stage is deep and irreversible. Histiocytes from granulomas which contain lymphoid cells, leucocytes, and crdiohistiocytes. Rheumatic granulomas would be usually located in perivascular connective tissue of the myocardium, in the endocardium, and (in slightly modified form) in synovial membrane of the articular bursa, periarticular and peritonsillar tissue, vascular adventitia, etc. At the phase of sclerosis, a cicatrix is gradually formed, which may develop at the site of fibrinoid changes and also as a result of cicatrization of rheumatic granulomas. Each phase of rhemuatism lasts on the average one or two months, and the entire cycle continues for at least six months.

Tissues in the zone of old cicatrices may be affected in relapses of rheumatism and sclerosis may develop. Affection of connective tissue of the valvular endocardium which causes sclerosis and deformation of the valve cusps (their adhesion to one another) is the most common cause of heart disease, while relapses of rheumatism (recurring attacks) aggravate affections of the valves. Non-specific exudative reactions may develop, mostly in the pericardium and the joints, and less frequently in the pleura, peritoneum, and the myocardium. Vasculitis (capillaritis, arteritis, phlebitis) are also non-specific rheumatic affections.

Clinical picture

The clinical picture of rheumatism is quite varied and depends mostly on the localization of the inflammatory changes in connective tissues of various organs and on acuity of the rheumatic process. As a rule, the disease develops in 1-2 weeks after a streptococcal infection (e.g. tonsillitis, pharyngitis or scarlet fever). Most patients develop subfebrile temperature, weakness, and hidrosis. Later (in 1-3 weeks), new symptoms develop to indicate affection of the heart. The patient complains of palpitation and intermissions in the work of the heart, the feeling of heaviness or pain in the heart, and dyspnoea.

Less frequently the onset of the disease is acute. Remitting temperature develops (38–39°C), which is accompanied by general asthenia, fatigue, and perspiration. Simultaneously (or several days later) the patient feels pain in the joints, (mostly in large joints, such as the ankle, knee, shoulder elbow joints, and in hands and feet). Affections of the joints are usually multiple and symmetric. The migrating character of pain is also characteristic: pain disappears in one joint and develops in others. Rheumatic polyarthritis is usually benign. Acute inflammation subsides in a few days, although dull pain in the joints may persist for a long time. Abatement of inflammation in the joints does not mean recovery because other organs become involved in the pathological process. The cardiovascular system is mostly involved, but the skin, serous membranes, lungs, liver, kidneys, and the nervous system may also be affected.

Examination of the patient with active rheumatism reveals pallid skin (even at elevated temperature) and increased perspiration. In some patients, the skin of the chest, neck, abdomen, and the face is affected by annular erythema (pale-pink painless rings not elevating over the surrounding skin). In other cases nodular erythema develops: circumscribed indurated dark red foci on the skin varying in size from a pea to plum; they are usually found on the lower limbs. If permeability of capillaries is increased, small haemorrhages into he skin sometimes occur. In rare cases, rheumatic subcutaneous nodules (firm, painless formations varying in size from a millet grain to beam) can be palpated, mostly on the extensor surfaces of the joints, along the course of tendons, and in the occipital region. The involved joints are swollen and oedematous, the overlying skin reddens and becomes hot to the touch. Articulation becomes very limited.

Lungs are affected in very rare cases. This is specific rheumatic pneumonia. Dry pleurisy or pleurisy with effusion are more common.

Heart affections may be the only clinical manifestation of rheumatism. On affected (rheumatic myocarditis). Rheumatic myocarditis is characterized by dyspnoea, the feeling of heaviness and pain in the heart, palpitation, and intermissions in the heart work. In addition, certain objective signs are found: enlargement of the heart, palpitation, and intermissions in the heart work. In addition, certain objective signs are found: enlargement of the heart, decreased heart sounds (especially the first sound); gallop rhythms develop in severe affection of the myocardium. A soft systolic murmur can be heard at the heart apes. It is associated with relative incompetence of the valve or affection fo the paillary muscles. The pulse is small and soft; tachycardia and arrhythmia are frequent. Arterial pressure is usually decreased. Circulatory insufficiency rapidly develops in grave diffuse myocarditis. Myocardial cardiosclerosis develops in benign outcome of the disease.

Rheumatic myocarditis usually concurs with rheumatic endocarditis (rheumocarditis). Early endocarditis signs are not pronounced (symptoms of myocarditis prevail). Further development of the heart disease proves the presence of endocarditis. At earlier stages of endocarditis systolic murmurs are coarser than in myocarditis; the murmur becomes louder after exercise; in some cases it becomes "musical". Diastolic murmur may be heard as well. It is probably explained by deposition of thrombotic mass on the valve cusps which produces turbulence in the blood flow as it passes from the atrium to the ventricle. These thrombotic deposits on the valves can leave their seat and become the cause of embolism or infarctions in various organs (e.g. the kidneys or the spleen). The mitral valve is mostly affected in endocarditis. Next in incidence follows the aortic valve; the tricuspid valve is affected still less frequently. If early attacks of rheumatic endocarditis are treated timely, development of the valvular heart disease may be prevented.

In a grave course of rheumatism the affection of the myocardium and endocardium may combine with rheumatic pericarditis, i.e. all membranes of the heart may be involved (pancarditis). Pericarditis may be dry or exudative.

The alimentary system is rarely affected. Acute pain in the abdomen (the abdominal syndrome) associated with rheumatic peritonitis (mostly in children) sometimes occurs. The liver is affected in certain cases (rheumatic hepatitis).

Affections of the kidneys are also common. Protein or red blood cells can be found in the urine due to affections of the renal vessels and (less frequently) developing nephritis.

The nervous system is often involved in rheumatism. This is due to either rheumatic vasculitis (attended by small haemorrhages or thrombosis of cerebral vessels) or inflammation of the brain and the spinal cord. Children would develop encephalitis with predominant localization in the subcortical nodes (chorea minor). It is manifested by emotional lability and hyperkinesia (abnormal movements of the extremities, the trunk, and the facial muscles).

Special laboratory tests help diagnose rheumatism. Moderate leucocytosis (with a shift to the left) is characteristic of acute rheumatism; eosinophillia, mono- and lymphocytosis may further develop. The erythrocyte sedimentation rate is always increased (to 50-70 mm/h in grave cases). Dysproteinaemia is characteristic: the albumin content drops below 50 per cent, the globulin content increases, and the albumin-globulin factor decreases below unity. A protein ogran shows increased a_2- globulin and g- globulin fractions; fibrinogen content increases to 0.6–1 per cent (normally it does not exceed 0.4 per cent). The blood contains C-reactive protein which is absent in healthy individuals. The level of mucoproteins increased and it can be revealed by a diphenylamine test. The titres of antistreptolysine O, antistreptohyaluronidase, and antistreptokinase increase significantly.

ECG often shows deranged conduction, especially atriventricular block of the first and second degree, extrasystole or other rhythm disorders, and decreased voltage of the ECG waves. Defective nutrition of the myocardium due to its inflammation can change the T wave and lower the S-T segment. Phonocardiograms show specific rheumocarditic changes in heart sounds, the appearance of murmurs, etc.

Course

An active rheumatic process continues for three to six months; in some cases it may be longer. Depending on the strength of the clinical symptoms and the character of the disease course, three degrees of rheumatic activity are distinguished: (1) maximum active (Acute) process with continuous relapses; (2) moderately active or subacute; and (3) rheumatism with minimal activity (flaccid or latent). If the clinical symptoms of the disease are absent and no signs of active rheumatism are revealed by laboratory testing, rheumatism are revealed by laboratory testing, rheumatism is considered inactive.

Rheumatism is characterized by relapses (recurring attacks) which are provoked by infection, overcooling, and physical overstrain. The clinical symptoms of relapses resemble the primary attack of the disease, but the signs of affection of the joints or serous membranes are less pronounced. Symptoms of heart affection prevail.

Treatment

Active rheumatism is treated at hospital. The patient should stay in bed and be given desentilizing and anti-inflammatory preparations, cortiocosteroid hormones, acetylsalicylic acid, amidopyrin, butadion, and chloroquine. Aantibiotics are given as well, especially in the presence of infectious foci (e.g. carious teeth, tonsillitis or sinusitis). These foci should be eliminated by appropriate means.

Prophylactic measures against rheumatism include hardening of the body, improvement of housing and working conditions, and control of streptococcus infection (treatment of chronic tonsillitis, otitis, sinusitis, etc.). Rheumatic patients should be kept under regular medical observation. To prevent relapses of rheumatism in spring and autumn, medicamentous prophylaxis should be given (bicillin in combination with salicylates or amidopyrine and chloroquine)..

BACTERIAL (SEPTIC) ENDOCARDITIS

Bacterial endocarditis is a severe general disease characterized by inflammation of the endocardium and ulceration of the heart valves in the presence of sepsis.

Acute bacterial endocarditis (a manifestation of acute sepsis) And subacute bacterial endocarditis are distinguished according to the course of the disease.

Subacute Bacterial Endocarditis

Aetiolgy and pathologenesis

Subacute bacterial endocarditis (endocarditis septica lenta) is usually caused by Streptococcus viridans, less frequently by enterococcus and Staphylococcus albus or St. aureus. In most cases, protracted bacterial endocarditis affects the valves that are changed by the rheumatic process, or develops in congenital heart diseases. It is believed that upset haemodynamics in heart diseases inflicts microinjuries to the valves which provoke changes in the endocardium, especially along the contact line of valve cusps. Factors weakening immunological reactions facilitate the development of the disease.

Pathological anatomy

This is characterized by the presence of ulcerous endocarditis. Ulcereated surfaces become covered with polyp-like thrombotic mass which sometimes looks like cauliflower. The valves become sclerosed and disfigured. The aortic valve is especially frequently involved. Endothelium of fine vessels is affected to cause vasculitis or thrombovasculitis: vascular permeability increases and small haemorrhages develop in the skin and mucosa.

Clinical picture

The symptoms of the disease mainly depend on toxaemia and bacteriaemia. The patients complain of weakness, rapid fatigue, and dyspnoea. As a rule, subfebrile fever first develops, which is followed by irregular elevation of temperature to 39 °C and more. Chills and excess sweating are characteristic. The skin and visible mucosa are pallid due to anaemia and aortic incompetence, which is characteristic of this disease. The skin sometimes becomes yellowish-grey (coffee with milk). Small haemorrhages in the skin, mucosa of the mouth (especially the soft and hard palate), on conjunctiva, and the eyelid folds (the Libman-Lukin symptom) indicate affection of the joints. Positive Konchalovsky-Rumpel-Leede sign is another indication of this process: if the arm of the patient is compressed by a tourniquet or by a cuff of a Riva-Rocci sphygmomanometer, multiple petechiae appear on the flexor surface of the elbow, and also distally of it. Brittleness of capillaries can also be established by pinching the skin. In most cases the patient's fingers become clubbed (Hippocratic fingers), while the nails are flat like a watch glass.

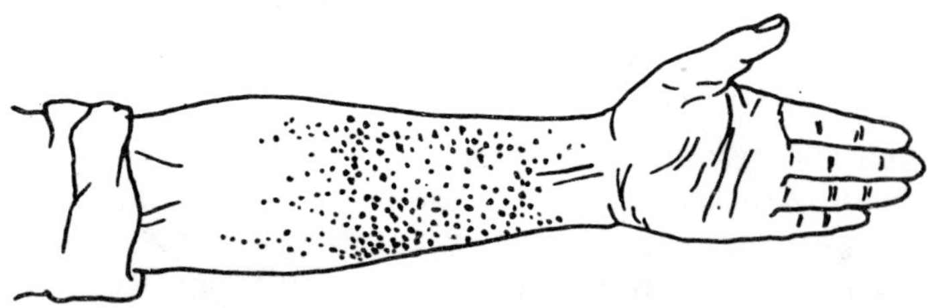

Fig. 2.17. Konchalovsky-Rumpel-Leede symptom.

Ausculation of the heart reveals signs of acquired or congenital diseases in most patients. Development of endocarditis is attended by the appearance of functional murmurs due to anaemia and murmurs that are caused by changes in the affected valve. Aortic valve cusps are usually involved and signs of aortic incompetence therefore develop, while symptoms of mitral insufficiency develop in affections of the mitral valve.

Subacute bacterial endocarditis is characterized by embolism (caused by decomposing thrombotic deposits on the valves) in the vessels of the spleen, kidneys or brain, followed by infarction of the involved organ. The spleen is enlarged due to the response of the memsenchyma to sepsis. The disease is characterized by involvement of the kidneys which gives the picture of diffuse glomerulonephritis and much less frequently focal glomerulonephritis without pronounced symptoms (slight proteinuria and haematuria, and insignificant cylindruria). Hypochromic anaemia (caused by increased haemolysis and inhibited erythropoiesis) and a markedly increased ESR are also characteristic. Leukocyte count varies. Eosinophil count decreases and there is a tendency to monocytosis and histiocytosis. If the number of histiocytes in blood taken from the ear lobe after its massage increases compared with their amount in the blood taken before the massage, it indicates affection of the vascular endothelium. Biochemical study of the blood reveals dysprteinaemia (hypoalbuminaemia, increased content of gamma-globulins); thymol and formal tests are positive. The causative agent of the disease can be revealed in blood cultures.

Echocardiography can be used to reveal vegetation on the cusps of the aortic valve, and less frequently on the mitral valve.

Treatment

The patient should be hospitalized and given antiboitics in large doses.

HEART VALVULAR DISEASES

Stable pathological changes in the structure of the heart that interfere with its normal function are called heart disease. Congenital and acquired diseases of the heart are distinguished. The incidence of acquired heart diseases is much higher.

Congenital diseases of the heart arise due to abnormal development of the heart and the great vessels during the intrauterine growth of the foetus with preservation of the intraterine character of circulation after birth. In defective division of the primary single-chamber arterial trunk into the pulmonary trunk and the oarta, and during formation of the heart chambers, defects in the interatrial and interventricular septa may be formed along with various abnormalities in the arrangement of the great vessels and their narrowing. Preservation of the intraauterine character of circulation after birth is the cause of patent ductus arterious (Botallo's duct) and patent foramen ovale. Congenital heart defects may often combine with communicated greater and lesser circulation systems and stenosis of the great vessels. Moreover, the valves (bicuspid, tricuspid, aortic, and pulmonary valves) may also have congenital defects.

Endocarditis, and especially rhemuatic endocarditis, is the main cause of acquired heart defects. Less frequently heart disease is the result of sepsis, atherosclerosis, syphilis, injuries, etc. Inflammatory processes occurring in the valve cusps often end in their sclerosis: deformation and shortening. An affected valve does not close completely to cause valvular incompetence. The cusps of the valves may adhere to one another because of inflammation to narrow the orifice they close. This narrowing is called stenosis.

Mitral Incompetence

Incompetenc to the mitral (bicuspid) valve (mitral insuffiieny) is in the complete closure of the atrioventricular orifice during life ventricular systole. As a result, the blood is regurgitated from the ventricle back to the atrium. Mitral incompetene may be organic and functional.

Organic insufficiency arisese as a result of rheumatic endocartiditis. Connective tissue develops in the cusps of the mitral valve which then contracts to shorten the cusps and the tendnos. The edges of the effected valve do not meet during systole and part of the blood is requrgitted through the slit into the left atrium from the ventricular during its contractin.

In functional (relative) incompetence the mitral valve is not altered but the orifice, which it has to close, is enlarged and the cusps fail to close it completely. Functional incompetence of the miral valve may develop because of dilatation of the left ventricle in myocarditis, myocardial dystrophy, or cardiosclerosis) and weakening of the circular muscle fibres that form the ring round the atrivoventricular orifrice. Affection of papillary muscles may also cause functional mitral incompetence. Functional insufficiency thus depends on dysfunction of the muscles responsible for the closure of the valve.

Haeodynaics

If the itral valve fails to close copletely during systole of the left ventricle, part of the blood is requrgitated into the left atrium. Blood filling of the atrium thus increases (because of the blood from the pulmonary veins which is added to the normal blood volume). Pressure in the left atrium increases, the atrium is dilated and becoems hypertrophied.

The amount of blood that is delivere into the left ventricle from the overfilled atrium during diastole exceeds normal and the atrium is thus overfilled and distended. The left ventricle has to perform excess work and becomes hypertrophied. Intensified work of the leflt ventricle compensates for the mitral incompetence during a long time. When the contractile power of the left ventricular myocardium weakens, diastolic pressure in it increases and this in turn inreases pressure in the left atrium.

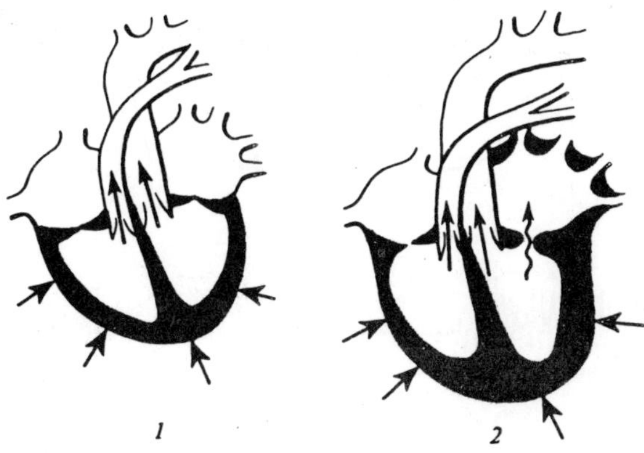

Fig. 2.18. Changes in the intracardiac haemodynamics in mitral incompetence.

Inccreased pressure in the left atrium increases pressure in the pulmonary veins and this in turn causes reflex contraction of the arterioles in the lesser circulation due to stimulation of baroreceptors. Spasm in the arterioles increases significantly pressure in the the pulmonary artery to intensify the load on the rigth ventricle which has to contract with a greater force in order to eject blood into the pulmonary trunl. The right ventricle can therefore also be hypertrophied during long-standing pronounced mitral incompetence.

Clinical picture

Most patients with mild or moderate mitral incompetence have no complaints for a long time and look very much like healthy subjects. As congestion in the lesser circulation develops, dyspnoea, palpitation of the heart, cyanosis, and other symptoms appear.

Papation of the heart area reveals displacement of the apex beat to the left and sometimes inferiorly. The beat becomes diffuse, intensified, and resistant,which indicates hypertrophy of the left ventricle. Percussion reveals displacement of the hearts borders to the left and superioroly because of the enlarged left atrium and left ventricle. The configuration of the heart becomes mitral with an indistint heart waist. The border of the heart shifts to the right in hypertrophy of the right ventricle. Ausculation of the heart reveals decreased first sound at the heart apex because the valves never close completely in this disease. Systolic murmur can be heart at the same point, which is the main sign of mitral incompetence. It arises during systole when the stream of blood passes a narrow slit leading from the left ventricle to the left atrium. The systolic murmur is synchronus with the first sound. When the blood pressure rises in the lesser circulation, an accent of the second sound can be heart over the pulmonary trunk.

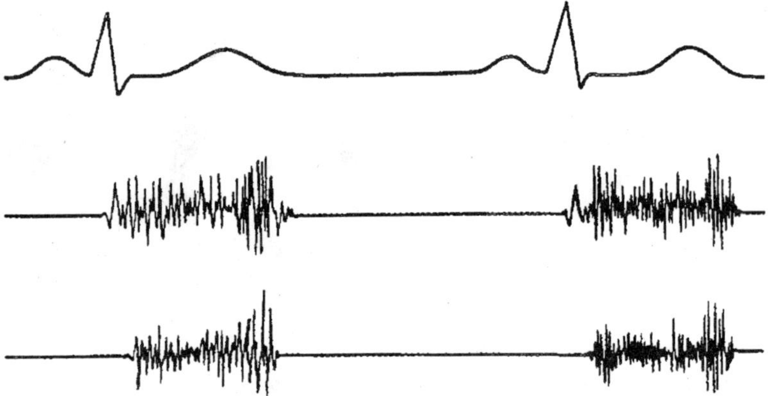

Fig. 2.19. PCG in mitral incompetence. The amplitude of the first sound is decreased on a PCG taken at the heart apex; systolic murmur occupies the entire pause between the first and second hdart sounds.

Ausculation findings are confirmed and verified by phonocardiography. The pulse and artterial pressure do not change in compensated mitral incompetence. X-ray studies show a specific enlargement of the left atrium and the left ventricle detectable by enlargement (to the left, superiorly and posteriorly) of the heart silhouette. When blood pressure increases in the lesser circulation, the pulmonay arch dilates. Signs of hypertrophy of the left atrium and the left ventricle can also be found on the ECG: it becomes the left type and the P waves become higher.

Mitral incompetence may remain compensated for a long time. But a long-standing pronounced mitral incompetence and decreased myocardial contractility of the left atrium and the left ventricle cause venous congestition in the lesser circulation. Contractility of the right ventricle can later be affected with subsequent development of congestion in the greater circulation.

Stenosis of the Left Atrioventricular Orifice

The left atrioventricular orifice usually narrows in a long-standing rheumatic endocarditis. In very rare cass mitral stenosis may be congenital or secondary to septic endocarditis. The atrioventricular orifice narows due to adhesion of the mitral cusps, their consolidation and thickening, and also shortening and thickening of the tendons. The valve thus becomes a diaphragm or a funnel with a slit in the middle. Cicatricial and inflammatory narrowing of the valvular ring isless important in genesis of mitral stenosis. The valve may be clacified in long-standing stenosis.

If stenosis is significant and the orifice is narrowed from the normal 4–6cm2 to 1.5 cm2 and less, haemodynamics becomes affected considerably. During diastole, blood fails to pass from the left atrium to the left ventricle and the remaining blood is added to the blood delivered from the pulmonary veins. The left ventricle thus becomes overfilled with blood, the pressure in the ventricle increases. Excess pressure is first compensated for by intensified contraction of atrium and its hypertrophy, but the force of the left atrial muscle is insufficient to compensate permanently for the pronounced narrowing of the mitral orifice and its contractile force soon weakens; the atrium becomes dilated, and the pressure inside it rises. This in turn increases pressure in the pulmonary veins, produces a reflex spasm in the arterioles of the lesser circulation and increases pressure in the pulmonary artery. All this requires intensified work of the right ventricle, which later also becomes hypertrohied. The left ventricle in mitral stenosis receives smaller volumes of blood and is therefore less active; its size slightly decreases.

Clinical picture

When congestive changes occur in the lesser circulation, the patient develops dyspnoea and palpitation on physical exertion; he complains of pain in the heart, cough, adn haemoptysis. Inspection reveals acrocy-anosis and cyantoic lush on the face. If the disease develops in childhood, the patient's physical growth often slows down and infantilism may develop ("mitral nanism"). Visual examination of the heart region often reveals a cardiac beat consequent upon dilatation and hypertroph of the right ventricle. The apex beat is not intensified; its palpation can reveal diastolic cat's purr (presystolic thrill). The broadening of cardiac dullness to the right and superiorly due to hypertrophy of the left atrium and right ventricle can be determined by percussion. The heart becomes "mitral" in configuration.

In auscultation of the heart the first sound at the apex becomes loud and snaping because the left ventricle receives little rblood and its contraction is fast. An adventitious sound due to the opening of the mitral valve can be heart at the apex beat. It follows the second sound of the heart. The loud first sound, second sound, and the sound of mitral valve opening give a specific murmur which is characteristic of mitral stenosis. The second sound becomes accentuated over the pulmonary trunk which pressure in the lesser circulation increases. Diastolic murmur is characteristic of mitral stenosis because the passage from the left atrium to the ventricle durign diastole is narrowed. This murmur can be heard to follow the mitral valve opening sound (protodiastolic murmur) because the velocity of the blood flow in early diastole is higher due to the pressure difference in the atrium and the ventricle. The murmur disappears when the pressure equalize. If stenosis is not pronounced, the murmur can e heard only at the end of diastole, immediately before systole proper (presystolic murmur); it arises during accelerationof theblood flow at the end of ventricular diastole because of the early atrial systole. Diastolic murmur can be heard in mitral stenosis during the entire diastole. It increases before systole and joins the first snapping sound.

Fig. 2.20. Changes in the intracardiac haemodynamics in mitral stenosis. The wavy arrow indicates an obstructed passage of blood from the atrium to the ventricle.

The puse in mitral stenosis may be different on the left and right arms. In considerable hypertrophy of the left atriu, the left subclavian artery is compressed and the pulse on the left arm becomes smaller (pulsus differens). If the left ventricle is not filled completely and the stroke volume is decreased, the ulse becomes small (pulsus parvus). Mitral stenosis is often complicated by atrial fibrillation, and the pulse becomes arrhythmic. Arterial pressure usually remains normal; the systolic pressure sometimes slightly decreases and diastolic pressure increases. X-ray pattern sof the heart show the specific enlargement of the left atrium, which leads to disappearance of the heart waist and "mitral" configuration appears. Enlargement of the left left atrium is determined in the first oblique position by the degree of displacement of the oesohagus which becomes especially vivid with barium sulphate suspension. If pressure in the lesser circulation increases, X-ray show swelling of the pulmonary arch and hypertrophy of the right ventricle. X-ray picture sometimes show calcification of the mitral valve. Pneumosclerosis develops during long-standing hypertension of the lesser circulation; it may also be revealed during X-ray examination.

The ECG of the heart with mitral stenosis shows hypertrophy of the left atrium and the right ventrile; the amplitude and duration of the P wave increase, especially in the first and second standard leads; the electrical axis of the heart deviates tothe right, a high R wave appears in the right chest leads and a pronounced S wave in the left chest leads.

A phonocardiogram taken at the apex shows the high amplitude of the first sound; the second sound is followed by the mitral valve opening sound and siastolic murmur; the amplitude of the second sound over the pulmonary artery increases compared with that over the aorta. If PCG and ECG are taken synchronously, attention should be paid to the lengthof the interval Q-1 sound (from the beginning of the Q wave on the ECG to the first soun on the PCG) and the second sound—OS interval. Echocardiograms in mitral stenosis are characterized by the following:

1. The A wave, describing the maximum opening of the valve during atrial systole either decreases or disappears altogether.

2. The speed of diastolic closure of the anterior mitral cusps decreases to decrease the E-F slope.

3. Movement of cusps change. The cusps of a normal mitral valve move in the opposite direction to set apart during diastole: the anterior cusps moves toward the anterior wall while the posterior cusp to the posterior wall. In stenosis, these movements become unidirectional because the more massive anterior wall pulls the posterior one by adhesion. The movement of the valve is represented on the echocardiogram in the form of a square wave. Enlargement of the left atrium adn changes in the cusps (fibrosis, calcinosis) can also be detected by echocardiography.

Mitral stenosis soon becomes attended by congestion in the lesser circulation which requires greater work of the right ventricle. Decreased contractility of the right ventricle and venous congestion in the greater circulation develop therefore in mitral stenosis earlier and more often than in mitral incompetence.

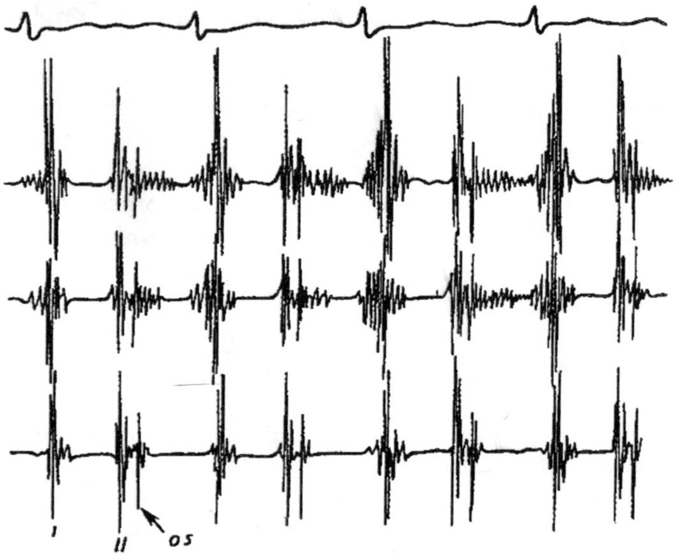

Fig. 2.21. PCG in mitral stenosis. The PCG taken at the heart apex shows increased amplitude of the first sound, diastolic (presystolic) murmur, and the sound of the mitral valve opening (OS).

Dilatation of the right ventricle and weakening of its myocardium are sometmes attended by relative tricuspid insufficiency. Moreover, long-standing venous congestion in the lesser circulation in mitral stenosis causes, wih time, sclerosis of the vessels and growth of connective tissue in the lungs. Another obstacle tothe blood flow is thus created in the lesser circulation and this adds to the difficulties in the work of the right ventricle.

Aortic Incometence

Aortic incometence (aoric insufficiency) is the failure of the aortic valve to clsoe comletely during ventricular diatole; blood thus leaks back intothe left ventricle. Aortic incometence is usualy secondary to rheumatic endocarditis, and less frequently bacterial (Setic) endocarditis, syhilitic affection of the aorta, or atherosclerosis. Inflammatory and sclerotic changes occuring in the base of the cusps during rheumatic endocarditis make them shrink and shorten. Atherosclerosis and syphilis can affect only the aorts (to distend it) while the valve cusps are only shortened. The cicatricial change may extend onto the cusps to disfigure them. Parts of the valve distintegrate in ulcerous endocarditis associated with sepsis and the cusps are affected with their subsequent cicartization and shortening.

Haemodynamics

During diastole, blood is delivered into the left ventricle not only from the left atrium but also form the aorts due to regurgitation, which overfills and sitends the left ventricle during diastole. During systole, the left ventricle has to contract with a greater force in order to expell the larger blood volume into the aorta. Intensified work of the left venticle causes its hypertrophy, while the increased systolic volume in the aorta cuases its dilatation.

Fig. 2.22. Changes in the intracardiac haemodynamics in aortic incompetence. The hypertrophied chambers are printed in red.

Aortic incompetence is characterized by a marked variation in the blood pressure in the aorta during systole and diastole. An increased volume of blood in the aorta during systole increases systolic pressure and since part of blood is returned during diastole into the ventricle, the diastolic pressure quickly drops.

Clinical picture

Subjective condition of patients with aortic incompetence may remain good for a long time becaluse the defect is compensated for by harder work of the powerful left ventricle; Pain in the heart (anginal in character) may sometimes be felt; it is due to relative coronary insufficiency because of pronounced hypertrophy of the myocardium and inadequate filling of the coronary arteries under low distolic pressure in the aorta. The patient may sometimes complain of giddiness which is the result of deranged blood supply to the brain (which is due to low diastolic pressure)

If contractility of the left-ventricular myocardium is impaired, congestion in the lesser circulation develops and the patient complains of dyspnoea, tachycardia, weakness, etc. The skin of the patient is pallid due to insufficient filling of the arterial system during diastole. Marked variations in the pressure in the arterial system during systole and diastole account for the appearance of some signs, such as pulsation of the peripheral arteries, the carotids (carotidshudder), subclavian, branchial, temporal, and other arteries; rhythmical movements of the head synchronous with the pulse (Musset's sign), rhythmial change in the colour of the nail bed under a slight pressure on the nail end, the so-called capillary pulse (Quincke's pulse), rhythmical reddening of the skin afte rubbing, etc.

The apex beat is almost always enlarged and shifted to the left and inferiorly. Sometimes, along with the elevation of the apex beat, a slight depression in the neighbouring intercostal spaces can be observed. The apex beat is palpable in the sixth and sometimes seventh hypochondrium, anteriorly of the midclavicular line. The apex beat isdiffuse, intense, and rising like a dome. This indicates significant enlargementof the left ventricle. The border of cardiac dullness can be found (by percussion) to shift to the left; the heart becomes "aortic" (with pronounced wais of the heart).

Ausculatation reveals decreased first sound at the apex, since during left-ventricular systole the period when the valves are closed is absent. The second sound on the aorta is also weak, and if the valve is damaged significantly, it can be inaudible. The second sound can be quite loud in atherosclerotic affection of the aorta. Diatolic murmur heart over the aorta and at the Botkin-Erb listening point is characteristic. This is a low blowing protodiastolic murmur which weakens by the end of diastole as the blood pressure in the aorta drops and the blood-flow rate decreases. The described changes in the sounds and murmurs are clearly visible on phonocardiograms. Murmurs of functional aetiology can be also be heard in aortic incompetence at the heart apex. If the left ventricle is markedly dilated, relative mitral incompetence develops and systolic murmur can be heart at the heart apex. Diastolic murmur (presystolic or Austin Flint murmur) can sometimes be heard. It arises due to an intense regurgitation of the blood that moves aside the mitral valve cusp to accoutn for functional mitral stenosis. Doubled sound (Trauble double sound) and doubled Vinogradov-Durozie murmur can sometimes be heart over the femoral artery in this disease.

The pulse in aortic incompetence is fast, full, and high, which is due due to high pulse pressure and increased volume of blood delivered into the aorta during systole. Arterial pressure constantly varies: the systolic pressure rises and siastolic falls, and the pulse pressure is therefore high.

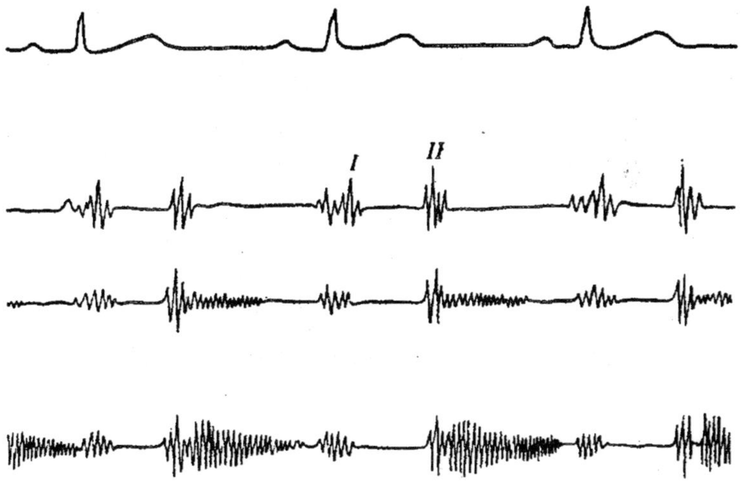

Fig. 2.23. PCG in aortic incompetence. The amplitude of the heart sounds and diastolic murmur are decreased on the PCG taken over the aorta.

X-ray studies show an enlarged left ventricle with a distinct waist of the heart and filatation of the aorta; pulsation of the aorta is intense.

The ECG also reveals various signs of hypertrophy of the left ventricle the electrical axis is deviated to the left, the S waves in the right chest leads are deep and the amplitude of the R wave is higher in the left chest leads; these signs often combine with signs of the overstrain in the left ventricle and relative coronary insufficiency (changes in the terminal part of the ventricular complex, displacement of the *S-T* interval, and the negative *T* wave).

Echocardiograms taken patients with aortic failure show flutter of the anterior mitral cusp during diastole caused by the thrust of the blood regurgitated from the aorta into the ventricle.

Aortic incompetence can for a long time be compensated for by internsified work of the hypertrophied left ventricle. When its contractile force decreases, congestion in the lesser circulation develops. Acute weakness of the left ventricle sometimes develops and is manifested by an attack of cardiac asthma. Dilatation of the weakened left venticle can cause relative mitral incompetence. This increases venous congestion in the lesser circulation associated with decompensated aortic incompetence and adds to the load on the right ventricle. This is mitralization of aortic incompetence, which may become the cause of venous congestion in the greater circulation.

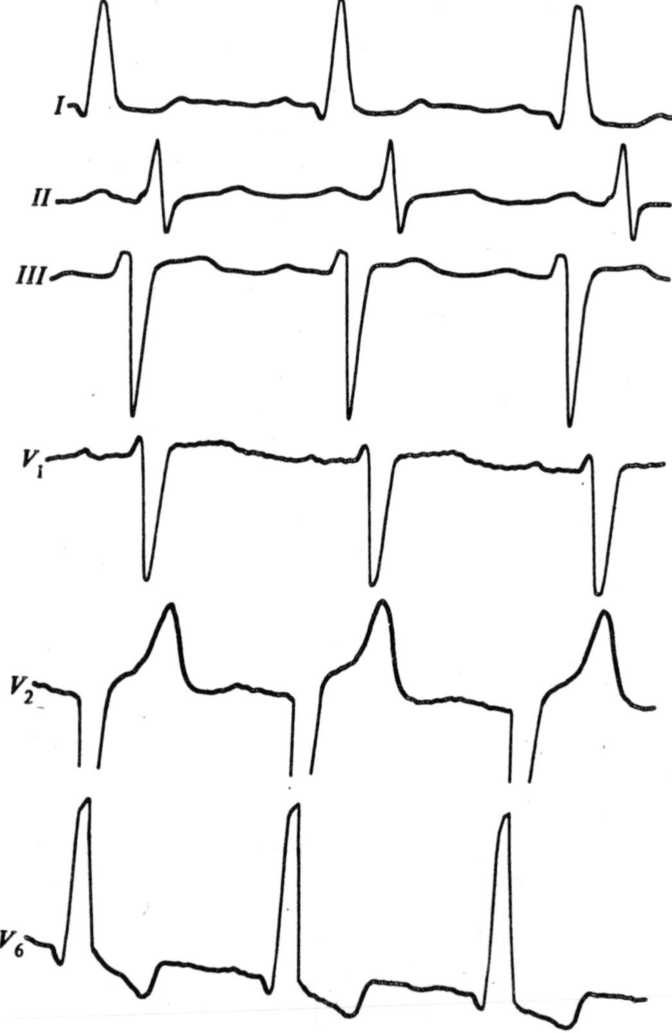

Fig. 2.24. ECG in aortic incompetence.

The narrowing of the aortic orifice (aortic stenosis) interferes with expulsion of blood into the aorta during contraction of the left ventricle. Aortic stenosis is usually caused by rheumatic endocarditis; less frequtnly it develops due to bacterial endocarditis, artherosclerosis, or it may be congenital. Setnosis results from adhred aortic valve cusps or develops due to cicatrical narrowing of the aortic orifice.

Haemodynamics

During systole, the left ventricle is not emptied completely because part of blood fails to pass the narrowed orifice into the aorta. A new normal portion of blood delivered during diastole from theleft atrium is mixed with the residual volume and the ventricle becomes overfilled. The presence inside it thus rises. This disorder is compensated for by an intensfied activity of the left ventricle to cause its hypertrophy.

Clinical picture

Aortic stenosis can remain compensated for yearsand would not cause any unplesant subjective sensations (even during intense physical exertion). If obstructon of the aortic orifice is considerable, insufficient blood ejection into the arterial system upsets normal blood supply to the hypertrophied myocardium and the patient feels pain in the heart (Angina pectoris-type pain). Disordered blood supply to the brain is manifested by giddiness, headache, and tendency to faintng. These symptoms like painin the heart would more likely occurs during physical exercise and emotional stress.

Fig. 2.25. Upset intracardiac haemodynamics in stenosed aortic orifice.

The skin of the patient is pallid due to insufficient blood supply to the arterial system. The apex beat is displaced to the left, less frequently inferiorly; it is diffuse, high, and resistant. Systolic thrill (cat's purr) can be palpated int he region of the heart. Percussion reveals displacement of the left border; the heart is "aortic" due to hypertrophy of the left ventricle. Auscultation of the heart at its apex reveals dimished first sound due to overfilling of the left ventricle and prolongation of systole. The second sound is diminshed over the aorta. If the aortic cusps adhere and are immobile, the second sound can be inaudible. Rough systolic murmur over the arota is characteristic. This murmur is generated by the blood flow through the narrowed orifice. It is conducted by the blood onto the carotids and can sometiems be heard in the interscapular space. The pulse issmall, slow, and rare, since the blood slowly passes into the aorta andits volume is decreased. Systolic arterial pressure is usually diminished, while diastolic remains normal or increases. The pulse pressure is therefore decresed.

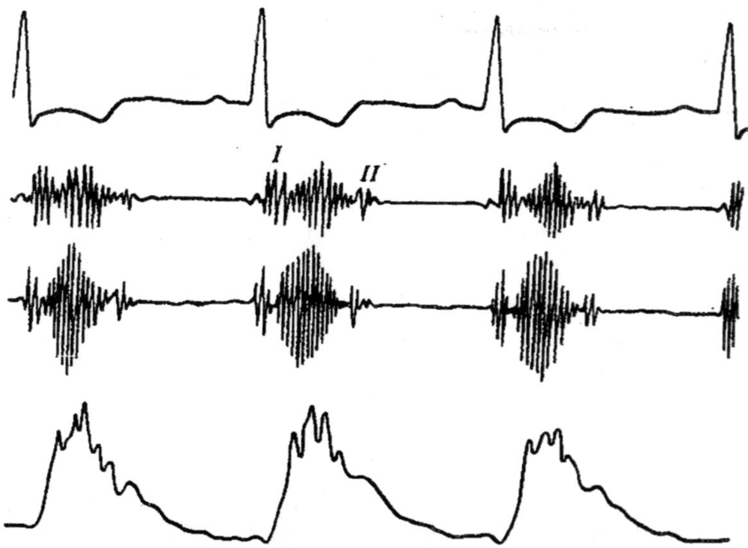

Fig. 2.26. PCG and sphygmogram in aortic stenosis. The PCG taken over the aorta shows crescendo-decrescendo (diamond-shaped) systolic murmur; the peak of the sphygmogram is serrated like a cock's comb.

X-ray examination shows hypertrophied left ventricle, "aortic configuration of the heart, and dilatation of the ascending aorta (post-stenotic); the cusps of the aortic valve are often calcified.

The ECG usually shows signs of hypertrophy of the left ventricle and sometimes of coronary insufficiency. The phonocardiogram shows the specific changes in the heart sounds: diminished amplitudes of the first sound at the heart apex and of the second sound over the aorta.

Systolic murmur over the aorta is typical; its oscillations are recorded in the form of specific diamond-shaped figures.

Sphygmograms of the cartoids reveal slowed ascent and descent of the pulse wave (slow pulse), small amplitude of the pulse waves, and the specific serrated pattern of their peaks (sphygmograms in the form of a cock's comb) showing oscillations associated with conduction of systlic murmur onto the neck vessels.

Echocardiograms show decreased opening of the aorticvalve during systole. Echoes from the cusps become more intense and signs of hypertrophyof the left ventricle appear.

Aortic stenosis remaisn compensated for a long time. Circulatory insufficiency develops in diminished contractility of the left ventrcle and it is manifested as in a aortic incompetence.

Tricuspid Incompetence

Insufficiency of the tricuspid valve can be functional (relative) and organic (permanent). Organic tricuspid incomptence occurs in rare cases, mainly due to rheumatic encocarditis. Tricuspid incompetence usually combines with affections of other heart valves; as an independent disease, it occurs in exceptionally rare cases.

Relative (functional) insufficiency of the tricuspid valve occurs more often. It is due to dialationof the rightr ventricle and distentionof the right-atrioventricular orifice. Tricuspid incompetence often combines with mitral diseases because the right ventricle has to perform greater work due to the high pressure in the lesser circulation. This causes overstrain and distention of the right ventricle.

Haemodynamics

Due to incomplete closure of the tricuspid valve durign right-ventricular systole, part of blood is regurgitated into the right atrium, where it is mixed with the normal volume of blood delivered from the venase cavae. The aterium thus becomes distended and hypertrophied. During diastole, a larger volume of blood is delivered into the right ventricule from the right atrium because the portion of blood that was regurgitated into the atrium during systole is added to the normal volume of blood delivered. This causes dilatation and hpertrophy of the right ventricle. Compensation in this disease isattained by intensified work of the right atrium and the right ventricle whose compensatory power is not great and congestion in the greater circulation therefore soon develops.

Clinical picture

Pronounced venous congestion in the greater circulation in the presence of tricuspid incompetence causes oedema, ascites, feeling of heaviness and right hypochondriac pain (due to enlargement of the liver). The skin becomes cyanotic, sometimes with a yellowish tint. The neck veins swell and pulsate; the positive venous pulse and pulsation of the liver are also observed. These pulsations are explained by the regurgitation of blood from the ventricle to the atrium trhoguh an incompletely closed satrioventricular orifice, owing to which pressure increases in theatrium and emptying of the neck and liver veins is made difficult. Tricuspid incompetence is characterized by extensive hepatic pulsation: the liver is displaced anteriorly and swells. This can be felt by trying to hold the liver by both hands: the pulsating liver sets the hands apart during the pulse wave.

Inspection of the patient reveals pronounced pulsation in the region of the right ventricle. As distinct from the heart beat in mitral stenosis, this pulsation is characterized by systolic retraction and siastolic protrusion of the chest.

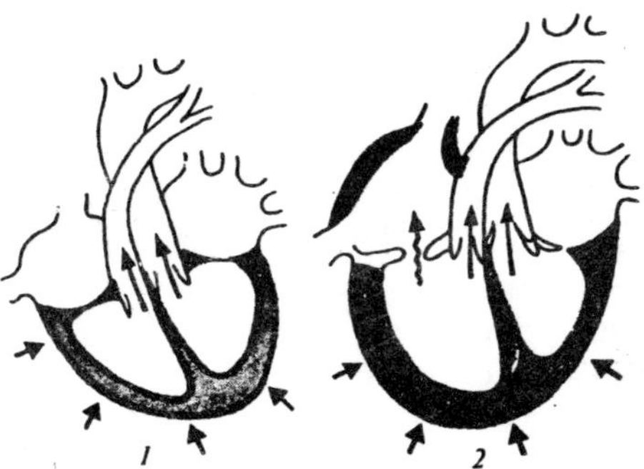

Fig. 2.27. Changes in intracardiac haemodynamics in tricuspid incompetence.

Systolic retraction of the chest in the region of the right ventricle is explained by pronounced diminution of its volume because much blood is delivered at that moment to thehepatic veins. Systolic reraction of the chest corresponds to the systolic swelling of the liver, and vice versa, diastolic overfilling of the right ventricle and protrusion of the chest in the ventricular region combine with diastolic diminution of the livervolume. Therefore, it the examiner places one hand on the region of the right ventricule and the other hand over the liver, he can feel specific rolling movements of the hands. The apex beat isas a rule not pronounced because the left ventricle is displaced posteriorly by the hypertrophied right ventricle. Percussion reveals marked displac-ement of the heart border to the right due to hypertrophy of theright atrium and right ventricle.

Auscultation at the base of the xiphoid process reveals diminished first sound; systolic murmur can be heart at the same listening point and also at the 3rd and 4th interspaces, to the right of the sternum; this murmur increases when the patient keeps his breath at the height of inspiration. Since pressure in the lessercirculation decreases in tricuspid incompetence, the second sound over the pulmonary trunk decreases in its clearness.

Pulse does not change significantly or it becomes small and fast, because serious heart failure often occurs in tricuspid incompetence. Arterial pressure usually decreases. The venous pressure increases markedly.

Signs of hypertrophy of the right heart's chambers can be detected roentegenographically. Electrocardiography also shows hypertrophy. Phonocardiography records systolic murmur at the base of the xiphoid process and at the 3rd and 4th intercostal spaces, to the right of the sternum; the systolic murmur has a decreasing character. When a phonocardiogram is taken at the height of inspiration, the vibration amplitude increases.

A phlebogram of the jugular vein reveals a high positive*a* wave which is connected with the intesnfied activity of the hypertrophied reight atrium, or the wave has the form characteristic of the positive venous pulse.

Echocardiography of the tricuspid valve is more difficult than that of the mitral of the aortic valve. Echocardiograms can reveal paradoxical movements of the interventricular septum in overloading of the right ventricle associated with tricuspid insufficiency.

Tricuspid incompetence usually combines with grave circulatory insufficiency. Long-standing congestion in the greater circulation upsets the function of many organs, sucha s the liver, kidneys, or the gastrointestinal tract. The liver is especially affected: prolonged congestion in the liver is attended by growth of connective tissue to provoke the development of the so-called 'cardiac fibrosis of the liver. This, in turn, even more interferes with the normal function of the organ and causes severe metabolic disorders.

Combined and Concomitant Heart Diseases

Acquierd diseases of the heart, rheumatic in particular, often occur as combined affections of the valves, i.e. valve incompetence and stenosis of the orifice occur simultaneously. Moreover, concomitant affection of two, and sometimes there valves (mitral, aortic, and tricuspid) may occur simultaneously.

Mitral incompetence is most common. It usually concurs with stenosis of the left venous orifice. Signs of both heart diseases are then found but one sign dominates as a rule; less frequently, signs of valvular incompetence adn stenosis are equally pronounced.

Dyspnoea and cyanosis are early symptoms of the mitral disease. The heart expands to the left, superiorly, and to the right, because both ventricles and the left atrium are hypertrophied. Intensity of the first sound at the apex depends on the prevalent disease: if mitral incompetence is the leading syndrome, the first heart sound diminishes; if mitral stenosis dominates, the first sound increases and becomes squelching. Two sounds are heard at the apex: systolic due to valvular insufficiency, and diastolic due to stenosed orifice.

Pulse and arterial pressure do not change in prevalence of mitral incompetence, while if mitral stenosis dominates, systolic arterial pressure may decreases and diastolic pressure increase; the pulse becomes small.

Combined aortic incompetence is usually secondary to rheumatic endocarditis. Both systolic and diastolic murmurs are characteristic; these sounds can be heard over the aorta. Vascular pulsation and high pulse pressure ar typical of aortic affection. At the same time, slow and small pulse, and low pulse pressure that are typical of aortic stenosis, are also less pronounced in combined aortic incompetence.

Detailed clinico-instrumental examination of patients with concomitant affections of several valves reveals signs typical of ech particular disease. It is necessary in such cases to conclude on the gravity of each disease and the prevalence of one of them. This is especially important for prognostic conclusions and for prospective surgical treatment.

Prognosis and Treatment of Heart Diseases

The course of heart diseases and their prognosis depend on many factors. Mild changes in the heart valves, in the absence of marked affections of the myocardium, can remain non-manifest for a long time without imparing the work capacity of the patient. Aortic insufficiency is compensated for a long time but when decompensation develops, the patient soon dies. Mitral stenosis has a worse prognosis because the disease is compensated by a weaker left atrium. Congestion soon develops in the lesser chirculation, which is followed by incompetence of the right heart chambers, and the greater circulation soon becomes affected by congestion. Repeated rhematic attacks have an adverse effect on the course of the diseases: the valvular apparatus of the heart is progressively ipaired and the myocrdium is affected. These disorders provoke circulatory insufficiency. Moreover, any infection, poisoning, physical or nervous overstrain, pregnancy and labour may give an impetus to the development of heart failure.

Restitution prognosis depends on the general condition and physical fitness of the patient. In the absence of symptoms of circulatory insufficiency, the patient may return tohis usual occupation; in the presence of signs of decompensation, the patient should be recommended to change his occupation. Work capacity can be preserved in a patient with heart disease provided the fulfills special recommendations for his work and rest, if he abstains from overeating, smoking, or drinking. Remedial exercises can be prescribed to strenghthen the myocardium in the absence of cardiac insufficiency.

Prophylaxis of heart diseses consists mainly in prevention of rheumatism, sepsis and syphilis. Prophylaxis also implies sanation of infectious foci, strengthening of the body, and exercise.

Conservative treatment of patients with heart diseases consists in prophylaxis and treatment of heart failure. Mitral stenosis is often corrected surgically (mitral commissurotomy). The adhered cusps are disjoined and the atrioventricular orifice is widened. This operation removes severe haemodynamic disorders. Aortic commissurotomy is performed in stenosed aortic orifice, but this operation is much more complicated. Surgical treatment of mitral and aortic incompetence consists in replacement of the destroyed valve by an artificial one.

Patients with rhemuatic heart disese should be regularly observed in out-patient conditions.

Myocarditis

Myocarditis is inflammation of the myocardium. The disease affects both men and women at any age. Acute, subacute, and chronic myocarditis are differentiated. The disease may be local or diffuse.

Aetiology and pathogenesis

Aetiological factors responsible for the development of myocarditis are various bacterial and virus infections. The most common cause of myocarditis is rhematism, next come sepsis, diphtheira, rickettsiosis, scarlet fever, and virus infections. Myocarditis may develop in sensitization to some medicinla preparations (allergic myocarditis). Inflammaoty changes in the myocardium caused by various infectiosn are the result of allergic reaction of the body sensitized by certain microbes. The microbe antigen or its toxin act on the mhocardium to cause formation of tissue antigens (auto-antigens) in it. As a result, auto-antibodies are formed which account for the vast change in the myocarditum.

Pathological anatomy

Dystrophic process in the muscle fibres are characteristic. Predominant exudative or proliferative processes are also observed in the interstitial tissue (in-terstitial myocarditis). The outcome of the inflammaoty changes is cardiosclerosis.

Clinical picture

This includes signs of decreased contractility of the myocardium and upset cardiac rhythm. The patient complains of dyspnoea in physical exertion, extreme weakness, palpitation, intermissions, dull and boring pain, or attacks of pain in the heart (like in angina pectoris).

The skin is pallid, sometimes with a slight cyanotic shade. In pronounced heart failure the neck veins become swollen. The pulse is small, soft, sometimes arrhythmical and accelerted: it may however be slowed down too. Extrasystole and, less frequently, paroxysmal fibrillation develop in deranged excitation function and automaticity.

Decreased diffuse apex beat, which is displaced anteriorly, is revealed on examination of the heart. Percussion can detect displacement of the heart to the left. Ausculation reveals a markedly decreased first sound at the early systole (due to decreased rateof rise in the intraventicular pressure). The second sound is eithe runchanged or diminished due to hypotension. Gallop rhythm can be heart in significantly decreased myocardial contractility. Systolic murmur can often be heard over the heart apex. It arises due to relative mitral incompetence. Arterial pressure, especially systolice, decreases and the pulse pressure falls accordingly.

ECG changes in myocarditis are quite varied and transient. Sinus tachycardia, sinus arrhythmia, and extrasystole (in the form of separate or group atrial or ventricular extrasystoles) can most frequently be found on electrocardiograms. Conduction is deranged according to incomplete or complete atrioventricular block. Diffuse affections of the myocardium are shown on ECG as diminished and split *P* wave, changed *QRS* complex (decreased voltage of the waves and their splitting), decreased *S-T* interval, the presence of two phases and inversion of the decreased *T* wave.

The blood counts show moderate neutrophilic leucocytosis with shift to the left, increased ESR, and hyperglobulinaemia (mainly due to a2- and g-globulins).

Course

The course of myocarditis is usually favourable and ends with recovery. Sclerosis of the myocardium develops in some patients (myocardial cardiosclerosis).

Treatment

Myocarditis patients are prescribed strict bed-rest. The main cause that provokes the disease is treated by antibiotics, desensitizing preparations, and hormones. Cardiac glycosides and diuretics are given in the presence of heart failure. In order to improve metabolism in the myocardium, carboxylase, ATP, vitamins, and potasssium preparations are prescribed. Deranged cardiac rhythm can be restored by appropriate preparations.

Cardiomyopathy

Cardiomyopathy includes disease of unknown aetiology characterized by affections of the heart muscle attended by enlargement of the heart and its insufficiency.

The following three types of cardiomyopathy would be usually differentiated: (1) congestive ordilatation; (2) hypertrophic; and (3) restrictive cardiomyopathy.

Congestive cardiomyopathy is characterized by dilatation of the heart chambers withpronounced heart failure. The patients develop dyspnoea durign slightest physical exertion and evenat rest, attacks of suffocation and cardiac pain which cannot be removed by nitroglycerin; heart palpitation and intermissions are also charactristic. As circulatory insufficiency progresses, the liver becomes enlarged and oedema and hydrops of the cavities develop. The borders ofthe heart are markedly displaced to the right, upwards, and to the left. The heart sounds at the apex are dulled, the second sound over the pulmonary trunk is accentuated, gallop rhythm often develops along with systolic murmur at the apex due to developong relative mitral insufficiency. The pulse is small and fast, sometimes arrhythmical. The arterial pressure is usually decreased.

ECG in congestive cardiomyopathy shows various changes in the myocardium; signs of overloading of the heart chambers and focal changes develop along with rhythm and conduction disorders.

Echocardiography reveals marked dilatation of the heart chambers and decreased contractility of the myocardium

Hypertrophic cardiomyopathy is characterized by the primary hypertrophy of the myocardium with subsequently developing cardiac insufficiency which differs but litter from congestive cardiomyopathy by its symptoms.

The special form of hypertrophic cardiomyopathy is obstructive cardiomyopathy, which is also known as diopathic hypertrophic subaortic stenosis. This cardiomyopathy is characterized by asymmetric hypertrophy of the interventricular septum in the region of blood outflow from the left ventricle. The cavity of the left ventricle diminishes. A circular ridge of the hypertrophied myocardium is formed beneath th eaortic valve. The ridge interferes with blood ejectioninto the aorta. This form of myocardiopathy is first manifested by symptoms characteristic for aortic stenosis: the patient develops headache, giddiness, faints, andheart pain (like in angina pectoris). Palpation and percussion of the heart reveals hypertrophy of the left ventricle; ausculation reveals coarse systolic murmur which is best heart at the 3rd and 4th intercostal spaceat the left edge of the sternum. The pulse is small and slow. Symptoms of circulatory insufficiency soon appear.

Echocardiography is most important among additional methods used to diagnose hypertrophic subaortic stenosis. It reveals asymmetrical hypertrophy of the interventricular septum, narrowing of the left ventricle and systolic deflection of the mitral valve cusp in the direction of the interventricular septum.

Hypertrophic cardiomyopathy canbe familial, inhereted by the autosome-dominant type.

Restrictive cardiomyopahtyis associated withdisordered distensibility of the myocardium due to endocardial and subendocardial fibrosis. The diastolicfunctionof the myocardium becomes upset andheart failure (without marked hypertrophy of the myocardium or dilationof the heart chambers) develops.

Pericarditis

Pericarditis is inflammationof the pericardium.

Aetiology and pathogenesis

In most cases pericarditis develops in the presence of rhemuatism ortuberculosis. Rheumatic pericarditisusually concurs with affection ofthe myocardiumand endocardium. Rhemuatic,andmostly tuberculous pericarditis are manifestationsof infectious allergic process. In certain cases, tuberculous pericarditis depends on the spread of infection from the foci inthe lungs and tracheobranchial lymph nodes (via the lymph ducts tothe pericarditis). Pericarditis can develop inother infectious as well(e.g. scarlet fever, measles, influenza, or sepsis). Sometimes itdevelops due to the transitionofinflammation fromthe adjacent organs in pleurisy, pneumonia, myocardial infarction, and also in injuriesto the heart and inuraemia.

Pathological anatomy

Dry (fibrinous) pericarditis and pericaditis with effusion are distinguished. Depending onthe character of effusion, pericarditis may be serous, serofibrinous, purulent, or haemorrhagic. Serous effusion and small masses of fibrinous exudate may be fully resorbed. Fibrinous and purulent effusion undergo organization to thicken the pericarditis membranes and cause commisures. Pericardium layers sometims become adherent and the pericardial sacdisappears (concretio pericardii). This condition is known as adherent (adhesive) pericaridits. Calcium is often deposited inthe affected pericardium and the heart becomes enclosed in a rigid case. The dense fibrous thickening ofthe pericardium compresses the heart. This condition is known as constrictive pericarditis.

Dry Pericarditis

Clinical picture

Pain inthe heart is often the only complaint of patients with dry pericarditis. The pain varies incharacter fromdiscomfor andpressure to strongtoruturing pain withradiationto the leftpart of the neck and the shoulder blade.

Inspection and percurssion of the heartdonotrevealany change unless myocarditis or other heart diseases concur. The most importantand sometimesthe only sign ofdry pericarditis isthe sound of pericardial friction. Dry pericarditis ends with complete recovery in 2-3 weeks; or the diseasemay convertinto pericarditis with effusion or adhesive pericarditis.

Pericarditis with Effusion

Clinical process

Patients complain of the pressing sensation in the chest and pain in theheart. AS effusion is accumulated, dyspnoea develops. Dysphagia dvelops in compressionof the oesophagus and hiccup when the phrenic nerve is compressed. Fever is an almost boligaroty symtom.

The appearance of the patient is characteristic: the face isoedematous and the skin is cyanotic and pallid. the neck veins are swollen due to an obstructed blood flowtotheheart viathe superior vena cava. Compressionofthis vein accounts for the oedematous appearance of the face, neck,and the anterior surface of thechest (Stokes' collar). The neck veinsmay sometimes swell only during expiration. If much exudate is collected in the pericardial sac, the patient assumes a characteris-ticposture: he sits inbed and inclines forward, his hands resting against the pillow lying onhis knees. The feeling of heaviness in the heart is thus lessened and respirtion made easier.

Inspection of the heart region reveals levelling of the interspaces. The apex beat is absent: if it is palpable, it appears to be displaced superiorly or medially of the left border. Percussion shows considerable enlargement of the cardiac dullness in all directions, absolute and relative dullness being almost undistinguishable. The area of dullness resembles a trapezium or a triangle: the right cardiohepatic angle becomes obtuse. If much exudate is accumulated,dullness extends to the left to diminish the tympany zone of Trabbe's space. Heart sounds are markedly decreased. The pulse is accelerated, small, and sometimes paradoxical. Arterial pressure is normal or decreased. Venous pressure is elevated. Palpation of the abdomen reveals marked enlargement of the liver due to congestion of blood in it. X-ray study shows enlarged heart silhouetee in the transvers direction and superiorly; the waste of the heart is absent, the pulsation is markedly weak, which is especially vivid on the X-ray picture.

The ECG shows low voltage of all waves and also change inthe S-T interval and the T wave inall standard leads. The S-T interval isfirstlocated above the isoelectric line and then below it. The T wave is first low and then becomes negative. The ECG changes are like those observed in myocardial infarctionexcept that they are equally pronounced in all leads (concordant), and that the Q wave remains unchanged.

Echocardiography is very important in thediagnosis of pericarditis with effusion. Echocardiography reveals the space between the pericardium and epicardium which is filled with fluid that does not returnthe probing ultrasound pulses. if effusion is small, th echo-free space can only be revealed between the pericardium and the posterior wall of the heart in a patientin the supine position. In the presence of a large effusion, this space can be revealed in the region of the anterior wall of the heart.

Course

Purulent pericarditis is a grave danger to the patient's life unless urgent treatment is given. Serous pericarditis may end incomplete recovery. Adhesive pericarditis produces persistent pathology because operative separation of the pleural layers is not sufficiently effective.

Treatment

Treatment depends on the cause of pathology. Rheumatic pericarditis is treated as rheumatism, and tuberculous pericarditis as tuberculosis. Various cardiacs and stimulatns should be given in the presence of cardiac insufficiency. If effusion is large, the pericarditium is punctured and the fluid withdrawn.

Essential Hypertension

Essentialhypertension (morbus hypertonicus) is the condition in which elevated arterial pressure is the leading symptom. The disease is provoked by nervous and functional disorders in the regulation of the vascular tone. Men and Women, mostly over 40, are equally attacked by the disease.

Essential hypertension shoudl accurately be differentiated from *symptomatic hypertension* in which arterial pressure rises as a symptomofsome other disease,this symptom being far from the leading one: Symptomatic hypertension occurs in aortic coarctation, artherosclerosis of the aorta and its large branches, in endocrine dysfunction(e.g. Itsenko-Cushing disease, phaeochromocytoma, primary aldosteronism, or the Conn syndrome), affection of the renal parenchyma, occlusive affection of the main renal arteries, and in some other diseases.

Aetiology and pathogenesis

Overstrain of the central nervous system, caused by prolonged and strong emotional stress and also mental overstrain, are believed to be the main cause of the disease. In some cases essential hypertension develops after brain concussion (concussion-commotion form). The importance of neurogenic factors was emphasized by Lang in 1922; later this hypothesis was confirmed by the Soviet physicians during World War II: the incidence of essential hypertension increaed significantly in the sieged Leningard.

Development of the disease greatly depends on occupation: it occurs mostly in subjects whose occupation is associated with nervous and mental overstrain, e.g. in scientific workers, engineers, physicians, drivers, etc. Familial preposition is another important factor.

The early stage of essential hypertension is characterized by nervous-functional disorder in regulation of the vascular tone. Vegetative-endocrine disorders and changes in the renal regulation of the vascular tone are later steps of the pathological process. Overstrain of the higher nervous activity causes vasopressor adrenal reaction by which arterioles, mainly the arterioles of the internal organs and especially of the kidneys, are narrowed. Ischaemia of the renin by the juxtaglomermular cells of the kidneys. Renin stimulates formation of angiotensin-II which in turn causes a pronounced pressor effect and stimulates secretion of aldosterone (sodium-retaining hormone) by adrenal cortex. Aldosterone promotes transition of sodium from the extracellular fluid into the intracellular fluid to increase the sodium content of the vascular wall. This causes oedematous swelling of the wall and narrowing of the vessel, which in turn promotes elevation of arterial pressure.

There is a system of depressor factors in the body whose dysfunction plays an important role in the pathogenesis of essential hypertension. Bradykinin and agiotensin, which produce a depressive effect, have been isolated. It is believed that the depressor system becomes altered by some unknown reason during essential hypertension.

Pathological anatomy

Essential hypertension gradually affects peremeability of vascular walls and their protein content. At late or grave froms of the disease, this causes sclerosis or necrosis of small arteries and secondary changes in the tissues of organs. Walls of large vessels are usually affected by atherosclerotic changes. The extent of vascular affection differs in various organs and various clinicoanatomical variants of the disease therefore arise, with a prevalent affection of the vessels of the heart, brain, or kidneys (primary cirrhosis of the kidneys thus develops).

Clinical process

During the early stage of the disease, the patient would usually complainof neurotic disorders: general weakness, impaired work capacity, inability to concentrate during work, deranged sleep, transientheadache, a feeling of heaviness in the head, vertigo, noise in the ears, and sometimes palpitation. Exertional dyspnoea (ascending upstairs, running) develops later.

The main objective sign of the disease is elevated systolic pressure (over 140–150 mm Hg) and siastolic pressure (over 90 mm Hg). Arterial pressure is very liable during the early stage of the disease, but later it stabilizes. Examination of the heart reveals signs of hypertrophy of the left ventricle: expanding apex beat, and displacement of the cardiac dullness to the left. The second sound is accentuated over the aorta. The pulse becomes firm and tense.

X-rays reveal "aortic" silhouette of the heart. The aorta is elongated, consolidated, and dilated.

The ECG is of the left type with displaced S-T segment, low,negative, or two-phase T wave in the 1st and 2ndn standard chest leads (V_5–V_6).

Cornoary atherosclerosisconcurs not infrequently. This may cause angina pectoris and myocardial infarction. In the late period of the disease, heart failure may develop due to fatigue of heart muscle as a result of increased arterial pressure. Heart failure is often manifested by acute attacks of cardiac asthma or oedema of the lungs; or chronic circulatory insufficiency may develop.

Vision may be deteriorated in grave cases. Examination of the fundus oculi reveals its general pallidness; the arteries are narrow and tortuous, the veins are mildy dilaed; haemorrhage into the retina (antiospasti cretinitis) is sometimes observed.

High arterial pressure in the affected cerebral vessels can derange cerebral circulation. This can cause paralysis, disorders in sensitivity, and sometimes deathof the patient. Cerebral circulation is deranged dueto spasms of the vessels, their thrombotic obstruction, haemorrhage due to rupture of the vessel, or diapedetic discharge of erythrocytes.

The affected kidneys become unable to concentrate urine (nicturia or isohyposthenuria develops). Metabolities (otherwise excreted with urine) are retained to provoke uraemia.

Essential hypertension is characterized by periodically recurring transient elevations of arterial pressure. Development of such crises is preceded by psychic traumas, ervous overstrain, variations in atmospheric pressure, etc.

Hypertensive crisis develops with a sudden elevation of the arterial pressure that canpersist from a few hours to several days. The crisis is attended by sharp headache, feeling of heat, persipiraton, palpitation, giddiness, piercing pain in the heart, sometimes by deranged vision, nausea, and vomiting. In severe crisis, thepatint mayu lose consciousness. The patient is excited, haunted by fears, or in indifferent, somnolent, and inhibited. Ausculation of the heart reveals accentuated second sound over the aorta, and also tachycardia. The pulse is accelerated but can remain unchanged or even decelerated; its tension increases. Arterial pressure increases significantly. ECG shows decreased S-T interval and flattening of the T wave. In the late stages of the disease, with organic changes in the vessels, cerebral circulation may be deranged during crisis; myocardial infarction and acute left-ventricular failure may also develop.

Classification

According to Myansnikov, three stages of the disease are classified; each stage is further divided into two phases, A and B.

The A phase of the first stage is latent; it is characterized by elevated arterial pressure durign a psychic stress, while under normal conditions arterial pressure is normal. The B- phase is transient (transitory hypertension); arterial pressure increases only occasionally and under certain conditions; objective changes are absent.

In the second stage, arterial pressure is elevatred permanently and more significantly. Phase A is characterized by permanent bu tunstable hypertension. Subjective sensations are pronounced; hypertensive crises are possible; spasms of the cerebral and coronary arteries are likely to occur as well. Signs of hypertrophy of the left ventricle develop. Phase B is characterized by a significant and stable elevation of the arterial pressure. Hypertensive crises are frequent. Paroxysms of angina pectoris and derangement of cerebral circulation of the angiospastic character occur. Changes in the fundus oculi and pronounced signs of hypertrophy of the left ventricle can be revealed.

During the third stage; sclerotic changes in the organs and tissues are observed along with stable and marked elevation of the arterial pressure. Phase A is compensated. Arteriosclerosis of the kidneys is observed, but the renal function is not upset significantly. Cardiosclerotic changes do not provoke stable heart failure;andsclerosis of cerebral vessels is not attended by pronounced disorders in the cerebral circulation. Phase B is decompensated, with grave dysfunction of various organs and with renal insufficiency; cerebral circulation is disordered and hypertensive retinophaty is observed. In this stage of the disease, the arterial pressure may normalize after infarction or apoplectic stroke.

Treatment

Complex therapy is required. Reasonable work should be alternated with rest, sufficient sleep, and remedial exercises. Sedatives should be given to improve sleep and to normalize excitation and inhibition processes. Hypotensive preparations (raauwolfia, ganglio blocking preparations, magnesium sulphate) are prescribed to inhibit the increased activity of the vasomotor centres and the synthesis of noradrenaline. Diuretics are given to decrease intracellular sodium; aldosterone blocking agents and other preparations are also given.

Atherosclerosis

Atherosclerosis is a chronic disese characterizd by systemic affection of arteries dueto metabolic disorders in the vascular wall. Atherosclerosis is one of the most common diseases and the most frequent cause of disability and premature death. It usually attacks people over 40–50 but sometimes occurs in younger patients. The incidence of atherosclerosis in men is 3–4 times higher than in women.

Aetiology

The aetiology of the disease is uncertain. It has been established that atherosclerosis occurs in subjects who undergo long nervous and pyschic overstrain, hypodynamia, who suffer from overeating (taking much food rich in animal fats) and obesity, and diabetes mellitus, or myxoedema. The pathogenesis of the disease is complicated and not clear. It is believed that the function of a complicated neurohumoral apparatus regulatign metabolism becomes upset by some causal factors. As a result biochemical, physio-chemical, and morphological composition of the blood is afected along with impairment of the blood coagulation system, disorders of strucure, biochemistry and function of the arteries, and of their permeability. Hereditary predisposition to atherosclerosis is important in some cases.

Pathological anatomy

The changes are localized in large elastic arteries, i.e. in the aorta, the coronary, cerebral, renal arteries and the large arteries of the limbs.

The atherosclerotic process occurs in stages. First lipids are deposited in the arterial intima to form fat spots and strips, whichdo not rise above the surface of the intima. Laer, connective tissue grows in the region of lipid accumulation to form a fibrous atherosclerotic plaque which rises above the intima surface to narrow the lumen of the vessel. The lipid-jprotein complex inside the plaque can decompose to ulcerate its surface and to form an atheromatous ulcer. The ulcerated plaques can become the cause of thrombosis of the vessel and calcifiction of the fibrous plaques, which causes even greater changes in the arterial walls.

Clinical picture

The disease may be asymptomatic for years (preclinical period). Study of the blood during this period shows increased cholesterol or beta-lipoprotein content. The further picture of the disease depends on the affection of particular vessels (aorta, coronary arteries of the heart, cerebral vessels, renal arteries, and thelimb arteries). This is the clinical period of the disease which is divied into three stages, namely the ischaemic stage, during which ischaemic changes occur in the organs (e.g. angina pectoris attacks occur in atherosclerosis of the coronary arteries); the second or thrombonecrotic stage, characterized by thrombosis in the changed arteries (myocardial infaraction may occur at this stage); and the third, fibrous stage, characterized by the development of connective tissue in the organs (e.g. cardiosclerosis).

Aortic atherosclerosis usually occurs in individuals aged over 40, but even its grave forms can sometimes be asymptomatic. Atherosclerosis of the ascendign aorta and its arch occurs more frequently. The patint feels pressing or burning pain in the retrosternal region which radiates in both arms, the neck, the back, adn the upper abdomen. As distinct from pain in angina pectoris this pain is more persistent; it can continue for hours and even day, with periodic weakening and strengthening. The elastic propertiesof the arota decrease and the heart has to perform greater work,whichcauses hypertrophy of the left-ventricular myocardium. Inspection of the heart reveals increased apex beat and its displacement to the left; on percussion, the borders of dullness of the vascular bundle are expanded. Retrosternal pulsation can be palpated in the jugular fossa because of the high standing of aortic arch (due to its elongation). The first heart sound is as a rule dull, the second sound isheart over the aorta; systolic murmur is also found whichappears or becomes intensified when the patient raises his arms. Systolic murmur is due to both sclerotic narrowing of the aortic orifice and roughness of the inner surface of the aorta. Maximum arterial pressure increases, while minimum pressure either remains unchanged or slightly falls. X-ray study reveals straightened, dilated, and elongated aorta. Atherosclerosis of the abdominal aorta is rarely diagnosed during life.

Atherosclerosis of the mesenteric arteries impairs blood suply to the intestine adn may cause attacks of angina abdominis: the patients suddently feels (3-6 hours after meals) piercing pain in the upper abdomen or in the ubrilical region, which persists from 2 or 20 minutes to 1-2 hours. The pain is attended by the swelling of the abdomen, regurgitation, constripation, palpitation, and increased arterial pressure. In thrombosis of mesenteric arteries, the intestinal loops begin necrotizing, which is manifested by haemorrhage and paralytic intestinal obstruction.

Atherosclerosis of the renal arteris causes vascular nephrosclerosis, whichis manifested by hpertension and isohyposthenuria.

Atherosclerosis of the cerebarl arteris is manifested by decreased work capacity (mental in particular), impaired memory, decreased active concentration, and rapid fagigue. The patient complains of insomnia and giddiness. The behaviour of the patient with pronounced atherosclerosis of the cerebral arteries changes: he becomes fussy, selfish, fidgety, and captious;p his mental power decreases. Atherosclerosis of the cerebral arteries is complicated by disturbed cerebral circulation (haemorrhage, thrombosis).

Atherosclerosis of the limb ateries. Clinical signs of this disease are pain in gastrocnemius muscle during walking. The pain is very severe and the patient has to stop for a while, but as he proceeds with walking, the pain soon develops again (intermittent claudication). The patient feels chill and cold in the legs. Examination of the legs reveals the absence ofpulse (or weak pulse0 on the dorsalis pedis artery and the posterior tibial artery. Dry gangrene of the lower extremities develops in grave cases due to local circulatory disorder. Angiograpy of the affected vessels reveals deformation, tortuosity, narrowing of the lumen, and microaneurysms.

Treatment

Combined therapy is aimed at removal of nervous effects, normalization of metabolic processes by prescribing rational diet and apprppriate medicinal preparatons to control the lipid content (lecithin, vitamin C, nicotinic acid, miscleron–2– capsules a day per os for several months, in courses, etc).

Rational diet, correct alternation of work and rest, regular exercise, quiet at home and at work are important conditions for prophylaxis of atherosclerosis.

Ischaemic Heart Disease

The term ischaemic heart disease includes many disease, such as angina pectoris, myocardial infarction, and coronary cardiosclerosis. The pathology is based on insufficient blood supply to the heart.The disproportion between the heart's demand for blood and the actual blood supply may arise when the heat's demands increase significantly, or when the actual blood supply diminishes for some reasons.

Angina Pectors

Anginapectoris is a frequently occurring disease. Its main clinical symptoms are attacksof retrosternal pain due to acute but transient disorder in the coronary circulation. The disease develops mainly in people over 40 and predominantly in men. Mental workers would mostly suffer from this disease.

Aetiology and pathogenesis

The most frequent cause of angina pectoris is atherosclerosis of the coronary arteries of the heart. Much less frequenlty it develops in infectious and infectious-allergic vascular diseases, such as syphilitic aortitis, panarteritis, periarteritis nodosa, rheumatic vasculitis, and obliterating endarteritis. Angina pectoris often cocurs with essential hypertension. Paroxysms of angina pectoris may develop due to upset nervous regulationof the coronary arteries as a reflex or in cholelithiasis, hiatus hernia, diseases of the stomach, etc. by reflex. Spasms of the coronary arteries (without their anatomical changes) can sometimes provoke angina pectoris. The spasm may develop in heavy smokers or as a result of astrong emotional stress.

Hypoxaemia (ischaemia) of the myocardium provokes the attack. Ischaemia develops in conditions when the insufficient amount of blood is delivered to the heart muscle through the coronary arteries, and the myocardium does not receive the necessary amount of oxygen. Transient oxygen hunger causes a reversible disorder in the oxidation-reducion processes in the myocardium. Stimulation of the interoceptors of the myocardium or the vascular adventitia by the products of upset metabolism produces a current of impulses via the centripetal pathways to the cerebral cortex to cause the specific symptom of the disease, retrosternal pain. High activation of the sympatheticoadrenal system is also very important for the onset of angina pectoris.

Pathological anatomy

No organic changes are sometimes foudn in those died from an attackofangina pectoris. But in 85-90 per cent of cases signs of atherosclerosis of the coronary arteries pronounced to a different degree are discovered.

Clinical picture

The main clinical symptom of the disease is pain in the centre of the sternum (retrosternal pain). Less frequently the pain originates in the heart region. The character of pain varies; the patient may feel constriction, compression, heaviness, burning, and sometimes sharp or stabbing pain. Pain is usually severe and the patient develops morid fear of death. Pain radiates into the left shoulder, left arm, left side of the neck and he head, the mandible, the interscapular space, and sometimes to the upper abdomen. The characteristic radiation of pain during angina pectoris isdue to hypersensitivity of the patient's skin to pain in the zones innervated by the 7th cervical and 1st-5th thoracic segments of the spinal cord. Stimuli from the heart are transmitted through these segments to the centrifugal nerves of the spinal cord (by the viscerosensory reflex). Pain arises under certain conditions: during walking (especially fast walking) and other exercises (angina pectoris of effort).

During physical exertion, the heart muscle requies more nutrition from blood. Atherosclerotic vessels cannot deliver the appropriate amount of blood and the patient has to stop exercise (walking) for a few minutes until pain subsides. Specific sign of angina pectoris is development of pain when the patient leaves a warm room and walks in the open during cold seasons. The symptom is even more pronounced if atmospheric pressure changes. Under emotional stress, the patient develops an attack of angina pectoris without any exercise. Attacks of pain may occur during sleep (angina pectoris at rest), after meals, in abdominal distention, and in high diaphragm. Pain may last from few seconds to 20-30 min. Quick removal of pain after taking nitrolycerin suggests angina pectoris.

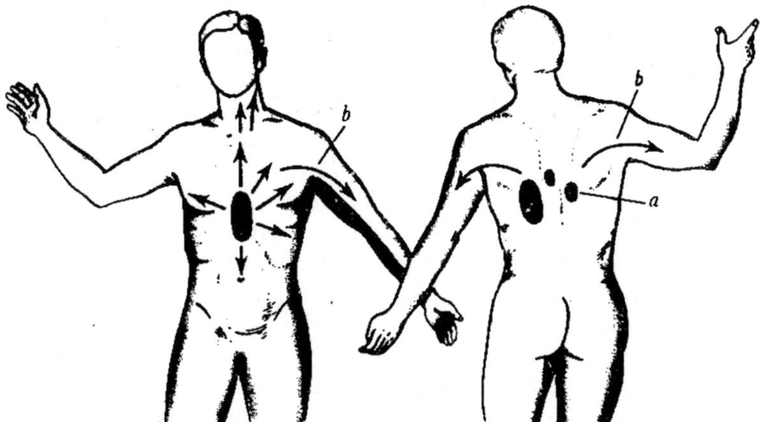

Fig. 2.28. Localization and radiation of pain in angina pectoris; Zakharyin-Head zone (a); nerve pathways of pain radiation (b).

The strength of attacks differs. In rare cases attacks end lethally. During an attack, thepulse isusually slow and rhythmical, but tachycardia, extrasystole, and increased arterial pressure are sometimes observed. Percussion and auscultation of the heart sometimescannot revealany abnormality, provided pronounced atheroscleroti cardioslecrosis is absent. The body temperture remains normal. Usually there are not changes in the peripheral blood.

Electrocardiographic studies during attacks ofangina pectoris sometimes reveal signs of disordered coronary circulation: the S-T segment is low, the two-phase or negative T wave is small in standard leads, and also in the corresponding chest leads, depending on the localizationof affection in the coronary system.When the attackisa-bated,the electrocar-diographic picture soon normalizes, ECG can sometimes reveal the described changes only during physical load. Coronarography with contrast substances is sometimes carried out to reveal occlusion of the coronary arteries.

Course

The disease is chronic. Attacks can be rare, onc a week or even less frequently; attacks may be absent for months or even years, or their frequency may increase and they become more severe. An attack of angina pectoris lasting more than 30-60 minutes can end in myocardial infarction. Patients with long-standing angina pectoris develop cardiosclerosis;the cardiac rhythm becomes disordered and symptoms of heart failure develop.

Treatment

Depending on the gravity of course, four functional classes of angina pectoris are distinghished. The first class is characterized by rare attacks and they occur only during physical exertion; in the second and third classes tolerance to exercise decreases progressively, while patients with the fourth class develop attacks of angina pectoris durign slight exercise or even at rest. Vasodilatory preparations are given during attacks of angina pectoris and measures taken to prevent new attacks. The mosteffective preparation to remove pain is nitroglycerin: 1–3 drops of a 1 per cent alcohol solution placed on a lump of sugar or a tablet (both should be kept under the tongue). The effectofnitr-oglycerin becomes appreciable usually in 1-2 minutes. The patient with angina pectoris should always keep nitroglycerin by himself. Mustard plaster or leeches on the heart are also effective.

In order to remove frequent and severe attacks ofpain and to prevent myocardial infarction, a surgical method of treatment is now used by whichthe coronary arteries are by-passed.

Sedatives (valerian, motherwort,trioxazin, etc.) in combination with vasodilatory preparations are regularly given to prevent attacks of angina pectoris.

Myocardial Infarction

Myocardial infarction is formation of a necrotic focus in the heart muscle due to upset coronary circulation. Myocardial infarction occurs mainly in people over 45; the incidence in men ishigher than in women.

Aetiology

One ofthemain causs of myocardial infarction (in at least 90–95 per cent cases) is atherosclerosis of the coronary arteries. In exceptionally rare cases, myocardial infarction is secondary to embolism of the coronary vessel in endocarditis or septic thrombophlebitis, in inflammatory affections of the coronary arteries such as rheumatic coronaritis, obliterating endarteritis, and nodular periarteritis. Overstrain, both physical and nervous, overeating, alcohol and nicotin poisoning can also provoke myocardial infarction.

Pathogenesis

The pathogenesis of myocardial infarotion is complicated and hasnot been sufficiently studied. According to current views, several factors are responsible for the onset of the disease. The main of them is believed to be coronary thrombosis and stenotic coronary sclerosis. Coronary thrombosis develops due to local changes in the vascular wall which are characteristic of atherosclerosis, and also as a result of disorders inthe blood coagulating system which are manifested by the decreased blood content of heparin and decreased fibrinolytic activity. In the absence of thrombosis, intense work of the heart in conditions of decreased blood supply to the myocardium (stenotic coronary sclerosis) is very important for the development of myocardial infarction. According toother investigators, the main factor responsible for myocardial infarction are functional disorders in the coronary circulation which provoke a prolonged spasm in the coronary circulation which provoke a prolonged spasm in the coronary arteries or their paresis. In their opinion atherosclerosis of the coronary arteries promotes development of myocardial infarction. Certain researchers believe that electrolyte imbalance in the heart muscle and accumulation of catecholamines in it and some other factors are important for the development of myocardial infarction.

Pathological anatomy

when blood supply to a part of the myocardium becomes inadequate, ischaemia develops, which is followed by necrosis. Inflammatory changes then develop round the necrotized area, and granular tissue grows. Necrotized mass isresorbed and replacaed by cicatricial tissue. The heart muscle may rupture at the site of necrosis with haemorrhage into the pericardialsac (heart tamponade). In gross myocardial infarction the scar tissue may be so thin that it can swell to give cardiac aneurysm. Myocardial infarction usually develops in the left ventricle. Muscle layers located beneath the endocardium are usually involved in the necrotic process (subendocardial form), but in severe cases, the entire muscle is involved (tansmural infarction). Fibrinous pericarditis usually arises in such cases. Fibrin is sometimes deposited on the inner membrane of the heart at sites corresponding to myocardial necrosis (parietal thromboendocarditis). Thrombotic masses may be torn off and carreid by the blood to accoutn for embolism of the cerebral, abdominal, lung, and other vessels. According to the size of the necrotized focus, micro- and macrofocal myocardial infarctions are differentiated.

Clinical picture

The outstanding Russian physicians Obraztosov and Strazhesko were the first to describe the clinical picture of myocardial infarction in 1909; they differentiated between three variatns of its course, namely, anginous, asthmatic, and abdominal (gastralgic) forms.

The angionous form occurs most frequently; clinically it is characterized by the pain syndrome. Pressing pain behind the sternum or in the region of the heart develops, like in angina pectoris. As a rule, pain radiates into the left shoulder and the left arm; less frequently into the right shoulder. Pain is sometimes so severe that cardiogenic shock develops which is characterized by the increasing weakness and adynamia, paleness of the skin, cold sweat, and decreased arterial pressure. As distinct from angina pectoris, pain in myocardial infarction is not removed by nitroglycerin and persists for longer gime (fro m30-60 minutes to several hours). Prolonged pain in myocardial infarction is termed as status anginosus.

The asthmatic form begins with an attack of cardiac asthma and lung oedema.The main syndrome is either absent or weak.

The abdominal form of myocardial infarction is characterized by pain in the abdomen, mostly in the epigastric region. The pain may be attended by nausea, vomiting, and constipation (gastralgic form of myocardial infarction). This form of the disease occurs mostly in farction of the posterior wall of the left ventricle.

Further observations have shown that the disease has considerably greater number of clinical signs. Myocardial infarction may sometimes begin with a sudden heart failure or collapse, various disorders in the cardiac rhythm or heart block, while the pain syndrome is either absent or is weak (painless form). This course often develops in recurrent infarction. The cerebral form of the disease is characterized by disorders in the cerebral circulation of various intensity.

Examination of the cardiovscular system reveals enlargement of cardiac dullness and low percussion sounds. The gallop rhythm can sometimes be heard. Pericardial friction is audible over a limited area, in the 3rd-4th interspaces, in transmural myocardial infarction. Pericardial friction becomes audible on the second or third day of the disease and persists for a few hours or may last one or two days. Pulse in myocardial infarction is often small, accelerted, or arrhythmical (in affection of the conduction system). Arterial pressure increases during path attacks but then it falls. Depending on the localizatio of infarction, circulation may be disordered by the left-ventricular or (less frequently) right-ventricular type. Inthe former case, congestive moist rales can be heard in the lungs; asphyxia resembling cardiac asthma may develop, which is then followed by oedema of the lungs. In the latter case, the heart is enlarged to the right; the liver is enlarged too; the lower extremities are affective by oedema.

Fever and leucocytosis develop on the second or third day of the diseae. They result form reactive processes which depend on absorption of the autolysis products from the site of infarction. The larger the necrotized area, the higher is the temperature and the longer the pyretic period and leucocytosis. Elevated temperature persists for 3-5 days, sometimes 10 days and more; ESR begins increasing and leucocytosis decreases from the second week of the disease.

Diagnosis of myocardial infarction depends substantially on the determination of activity of some blood serum enzymes which are released due to necrotic changes in the myocardium, e.g. the activity of creatine phosphokinase, the first and the fifth enzymes of lactic dehydrogenase, aminotransferase, and especially asparagine and (to a lesser degree) alanine increase by the end of the first day of the disease. The activity of creatine phosphokinase normalizes in 2-3 days, of aminotransferase in 4-5 days, and of lactic dehydrogense in 10-14 days.

Electrocardiographic examination is especially important. It establishes the presence of myocardial infarction and also some important details of the process such as localization, depth of the process, and the size of the affected area.

The *S-T* segment and *T* wave change during the first hours of the disease. The descending limbof the R-wave transforms into the *S-T* segment without reaching the the isoelectric line. The *S-T* segment rises above the isoelectric line to form a convexing arch and to coincide with T wave. A monophase cure is thus formed. These changes usually persist for 3–5 days. Then the *S-T* segment gradually lowers to the isoelectric line while the T wave becomes negative and deep. A deep Q wave appears, the R wave becomes low or disappears at all. The QS wave is then formed, whose apperance is characteristic of transmural infarction. Depending on localizationof infarction, changes in the ventricular complex are observed in the corresponding leads. The initial shape of ECG can be restored during cicatirization, or the changes may remain for the rest of life.

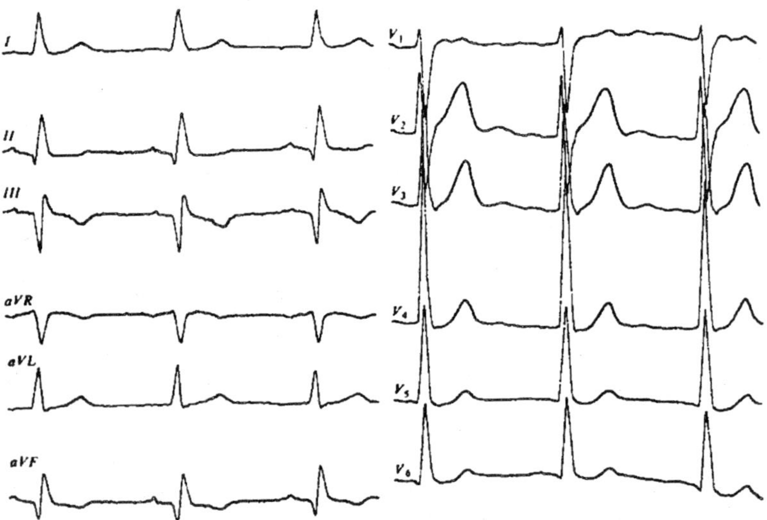

Fig. 2.29. ECG signs of focal myocardial infarction in the posterior wall of the hypertrophied left ventricle during cicatrization.

The radionucleide method can be helpful in diagnosing myocardial infarction. The patient is given pyrophosphate labelled with ^{99}Tc which is mainly concentratred in necrotized tissue, i.e. in the zone of infarction. The extent of the affection can thus be estimated.

Retrosternal opperssion and general weakness are the usual symptoms in microfocal myocardial infarction. Moderately high temperature usually persists for 1–2 days. Mild leucocytosis in only transient; ESR is slightly increased and the enzymatic activity is increased. The ECG changes are as follows: the S-T segment is below and sometimes above the isoelectric line, and the T wave is either negative or two-phase. These changes disappear in a few days or in a month.

Course

This depends on the size of the affected area, the condition of other arteries of the art and condition of other arteries of the heart and collateral circulation, on the degree of cardiac and circulatory insufficiency, and on the presence of complications. Cardiorrhexis is among them. It occurs during the first ten days of the disease, during pronounced myomalacia leading to rapid (within a few minutes) death. Fatal outcome may be caused by ventricular fibrillation. Cardiac aneurysm can develop during the disease. An acute aneurysm occurs during the first days of transmural infarction: the increased intraventricular blood pressure causes a protrusion at the site of myomalacia of intact layers of the heart wall. The aneurysm is usually formed in the wall of the left ventricle.

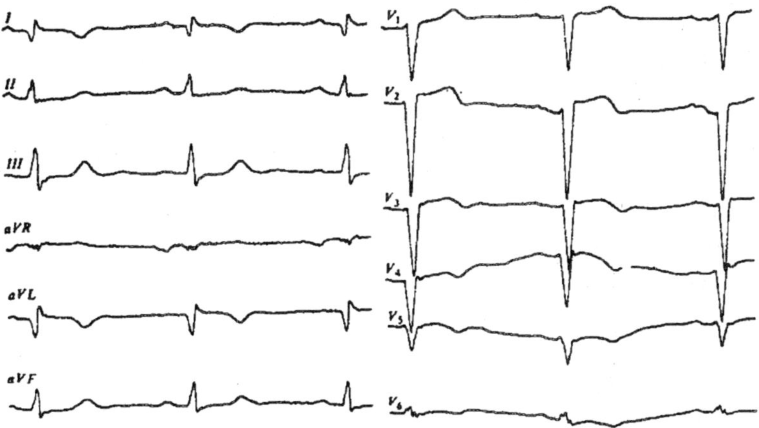

Fig. 2.30. Massive myocaridal infarction of the entire anterior wall of the left ventricle, the apex, and the lateral wall.

The clinical picture of actue cardiac aneurysm is characterized by pericardial pulsation in the 3rd or 4th intercostal space, to the left of the sternum. Ausculation of the heart reveals the gallop rhythm, systolic murmur, and pericardial friction. An acute cardiac aneurysm is covered into chronic when the necrotized site turns into a scar of connective tissue, or chronic aneurysm can develop at later periods of the disease. Signs of chronic aneurysm are pericardial pulsation, extension of the heart border to the left, systolic murmur in the region of the aneurysm, and a "stabilized" electrocardiogram on which the changes characteristic of acute period of the disease are perserved. X-ray examinations show enlargement of the heart silhouette and paradoxical pulsation. Chronic cardiac aneurysm causs heart failure which is difficult to cure.

Cicatrization of uncomplicated myocardial infarction continues for one to three months. The process ends in focal cardiosclerosis.

Treatment

Patients with myocardial infarction are hospitalized during the first hours of the diseae. Absolute rest is required. The diet restricted in calories should be free from food that may cause flatulence of the intestine. In the presence of pain attacks, nacrotics (morphine hydrochloride and others) are given subcutaneously together with atropine which removes possible side effects such as nausea and vomiting. If nacrotics do not help, nitrous oxide mixed with oxygen should be given to inhale through a special anaesthetic apparatus.

In cases with circulatory insufficiency, strophanthin should be given; noradrenaline or mesaton shculd be given intravenously by a drip method in collapse. Anticoagulants (of direct and then indirect action) should be given during the very first hours of the disease. The duration of hospital stay depends on the gravity of the clinical picture and the degree of the myocardial affection. Spasmolytics and sedatives should be given during the entire diseae. Gross transmural, and also recurrent myocardial infarctions should be treated for 3–6 months. The patient often remains disabled.

Cardiosclerosis

Cardiosclerosis (fibroid heart) is the disease of the myocardium caused by developing fibroid elements (cicatrix) in the heart muscle. Atherosclerotic and mycarditic cardioscleroses are distinguished. The later may result fromany myocarditis. Atherosclerotic cardiosclerosisis the result of atherosclerosis of the coronary arteries (diffuse atherosclerotic cardiosclerosis). Myocardial infarction cardiosclerosis.

Clinical picture

The patient complains of decreased work capacity, dyspnoea, first only during heavy exertion, and if the heart muscle is affected considerably, these symptoms appear during usual walk or work. Objective examination of the patient reveals enlargement of the cardiac dullness to theleft. The heart becomes "aortic", its sounds are dulled. Pronounced signs of heart failure may develop later. Cardiosclerosis is the most frequent cause of various arrhythmias such as extrasystole (usually ventricular), atrial fibrillation, or heart block.

Course

Since coronary atherosclerosis tends to progress, cardiosclerosis also becomes more pronounced with time. Myocarditi cardiosclerosis has no progressive tendency.

Treatment

It is the same as that usually given in atherosclerosis, angina pectoris, circulatory insufficiency, and heart arrhythmias.

3

DIGESTIVE SYSTEM

SPECIAL PATHOLOGY

Achalasia of the Cardia

Achalasia (syn. Cardiospasm, megaoesophagus) is characterized by the failure of atonic muscles of the cardia to relax, distrubed persistaltic activity of the esophagus, deranged reflex opening of the cardia on swallowing, and hence impaired passage of food into the stomach and dilation of the oesophagus due to retention of food in it.

The aetiology and pathogenesis of this condition are not sufficiently studied.

Pathological anatomy

The disease is characterized by dystrophic changes in the intramural nerve plexus of the oesophagus and cardia, dilation of the oesophagus, and congestive esophagitis.

Clinical picture

Restrosternal pain, dysphagia and regurgitation are the main symptoms. The pain manifests itself by paroxysmal attacks occuring usually at night, and is often associated with the empty or, on the contrary overdistended esophagus. Dysphagia is first occasional. In advanced cases it occurs during each meal and is especially marked during swallowing dry or inadequately chewed food in hurried eating. Emotional stress intensifies bysphagia. In order to facilitate the passage of food into the stomach, the patient resorts to various tricks such as drinking a gulp of water or taking a deep breath, which sometimes helps. The food mass is regurgitated as a person bends down with his oesophagus overfilled or during night sleep. The regurgitated mass may be aspired causing recurrent aspiration pneumonia.

The diagnosis is confirmed by X-ray examination which demonstrates dilatation and elongation of the esophagus (sometimes a sigmoid dilatation) of the varying degree, upset perstalsis, accumulation of liquid (saliva, mucus) or food in the oesophagus while the stomach is empty. A barium meal is retained in the oesophagus and its upper level is often as high as the clavicle. (the barium swallow would then suddenly 'fall' into the stomach). The cardial segment of the oesophagus is narrowed and has regular outlines; it does not open on swallowing and the barium meal does not pass into the stomach; the gastric air bubble is absent (the pathognomonic sign).

Course

The disease progress rapidly with intensifying dysphagia and other symptoms; weight loss is also progressive. Recurrent pneumonia and chronic bronchitis supervene due to aspiration of regurgitated oesophageal contents.

Treatment

The efficacy of medication is low. The condition is commonly corrected by cardiodilaltation using Einhorn's dilator, or by balloon dilatation which are usually performed in specialized hospitals. Cases with severe dilataton and elongation of the oesophagus require surgical treatment.

Oesophagitis

Oesophagitis is inflammation of the oesophagus. It is a common disease of the gastrointestinal tract. Acute and chronic oesophagitis are distinguished.

Aetology and pathogenesis

Acute oesophagitis develop in response to irritation of the oesophageal mucosa by hot food or drinks, by chemicals (iodine tincture, strong acids, or alkalies); it may be secondary to acute infectious diseases such as scarlet fever, diphtheria, sepsis, or may attend acute pharyngitis or gastritis. Chronic oesophagitis develops due to repeated exposure of the oesophageal mucosa to hot or rough (spicy) food, alcohol, some indistrial poisons that are inhaled with dust or ingested with food, etc. Congestive oesophagitis develops due to retention and decomposition of food in the oesophagus in achalasia and oesophageal stenosis patients. A more common cause of subacute and chronic oesophagitis is the reflux of the gastric juice into the oesophagus due to cardial insuffiency (reflux oesophagitis). Reflux oesophagitis usually occurs in axial hiatal hernia. Cardial insufficiency and reflux oesophagitis may also be secondary to surgical procedures (resection of or injury to the cardial sphincter). Relative, i.e. functional cardial insufficiency may be seen in peptic ulcer, cholelithiasis, and some other diseases. It is due to spastic contraction of the pylorus or gastric hypertonia and high intragastric pressure. Cardial insufficiency and reflux oesophagitis occur in pregnancy and large newgrowths of the abdominal cavity (due to high intra-abdominal pressure).

Pathologenesis

The condition arises as a result of irritation which may be thermal chemical, toxic, or peptic (in cardial insufficiency and reflux oesophagitis); bacterial toxic or toxic-allergic effects may in rare cases produce oesophagitis as well.

Clinical picture

This depends on the acuity of the process, its aetiology and extension. Acute catarrhal oesophatitis presents with pain on swallowing, discomfort, and sometimes dysphagia. Haemorrhaagic oesophagitis caan be manifested by haematemesis and melaena. Pseudomembranous oesophagitis (it usually supervenes diphtheria and scarlet fever) is characterized by the presence of fibrin membranes in the vomitus. The clinical picture is especially severe in cases where abscess and phlegmon of the oesophagus are aggravated by sepsis and toxaemia (phlegmonous oesophagitis).

Chronic oesophagitis present with heartburn and retrosternal discomfort. In rare cases there is pain and dysphagia. The main symptoms of reflux oesophagitis are heartburn and regurgitation which intensify on lying down or when the patient bends. Retrosternal pain on swallowing (sometimes resembling coronary pain) is not infrequent.

Radiography provides little evidence for the diagnosis of oesophagitis. In reflux oesophagitis it usually reveals hiatla hernia and visualizes gastro-oesophageal reflux. The patient must be examined in both the vertical and horizontal positions, and the intra-abdominal pressure must be raised by coughing, sneezing, straining the prelum, or by applying the pressure of the X-ray tubulus to the epigastrium. Oesophagoscopy is decesive in the diagnosis of oesophagitis. Using this technique allows to assess the degree of inflammation, its extension and character.

Complications

Phlegmon and abscess of the oesophagus may be complicated by its perforation and subsequent severe mediastinitis or peritonitis. Haemorrhagic and erosive oesophagitis may be attended with oesophageal bleeding. Severe acute and chronic oesophagitis may cause stricture and cicatricial shortening of the oesophagus which, in turn, promotes formation or enlargement of the existing axial hiatal hernia.

Treatment

Patients with acute corrosive oesophagitis and abscess and phlegmon of the oesophagus should be managed in a hospital setting. In all acute and subacute oesophagitis cases bland feeds are prescribed. Fasting is sometimes advisable for several days. Antibiotics are administered in abscess and phlegmon cases. Astringents (bismuth nitrate, 1.0 g, or 20 ml of a 0.06 per cent silver nitrate solution, 4-6 times a day before meals) are given for other acute forms of oesophagitis. Any job that might require bending down or strain of the abbominal muscles is contraindicated to a patients with reflux oesophagitis. Sleeping in the semireclining position is recommended.

STOMACH

Methods of Examination

Inquiry

Complaints

Patients with diseases of the stomach complain of poor appetite, perverted taste, regurgitation, heartburn, nausea, vomitting, epigastric pain, and haematemesis. Regurgitation, heartburn, nausea, vomitting, and the feeling of overfilled stomach after meals are the group of the s-called dyspeptic complaints. These symptoms may be observed in diseases of some othe rorgans and systems. Determining the specific character of each symptom is important during inquiry of the patient.

Deranged (poor or increased) appetite occurs in infectious diseases, metabolic disorders, etc. Poor appetite or its complete absence (anorexia) is usually characteristic of gastric cancer. This symptom is often an early sign of cancer. Appetite often increases in peptic ulcer, especially in duodenal ulcer. Loss of appetite should be differentiated from cases when the patient abstains from food for fear of pain (cibophobia). This condition often occurs in subjects with gastric ulcer, though their appetite is incresed.

Perverted appetite that sometimes occurs in patients is characterized by the desire to eat inedible materials such as charcoal, chalk, kerosine, etc. Appetite is perverted in pregnant women and in person suffering from achlorhydria. Some patients with cancer of the stomach or some othe rorgans often feel aversion to meat. The developmental mechanism of appetite is connected with excitation of the food centre (according to Pavlov). Excitation or inhibition of this centre depends on impulses arriving from the cerebral cortex, on the condition of the vegetative centres (excitation of the vomiting centre causes loss of appetite), and on reflex effects from the alimentary organs. The multitude of factors that act on the food centre accont for the high variation in appetite.

Taste may be perverted due to the presence of unpleasant taste in the mouth and partial loss of taste in an individual. It can often be associated with some pathology in the mouth, e.g. caries or chronic tonsilitis. A coated tongue can be another cause of unpleasant taste in the mouth.

Regurgitation usually implies two phenomena: a sudden and sometimes loud uprise of wind from the stomach or oesophagus (eructation), and the return of swallowed food into the mouth (sometimes together with air). Regurgitaton depends on contracton of the oesophageal muscles with the open cardia. Regurgitation may be due to air swallowing (aerophagy). It is heard at a distance and occurs in psychoneurosis. In the presence of motor dysfunction of the stoamch, fermentation and putrefaction of food with increased formation of gas occur in the stomach (the phenomenon otherwise absent in norm). In abnormal fermentation in the stomach, the erutated air is either odourless or smells of bitter oil, which is due to the presence of butyric, lactic and other organic acids that are produced during fermentation in the stomach. In the presence of abnormal putrefaction, the belched air has the odour of rotten eggs (hydrogen sulphide). Bitter belching indicates intensive degradation of proteins. Belching is characteristic of stenosed pylorus with great distention of the stomach and significant congestion in it. Acid regurgitation is usually associated with hyper secretion of gstric juice and occurs mostly during pain attacks in ulcer. But it can also occur in normal or insufficient secretion of the stomach in the presence of insufficiency of the cardia (when the stomach contents are gurgitated into the oesophagus). Bitter regurgitation occurs in case with belching up of bile into the stomach from the duodenum, and also in hyperchlorhydria; bitterness depends on the bitter taste of peptones.

Pyrosis is otherwise known as heartburn, i.e. burnin pain in the epigastric and retrosternal region. Heartburn arises in gastro-oesophageal reflux, mostly in the presence of gastric hypercidity in various diseases of the alimentary tract (e.g. peptic ulcer or cholecystitis), hiatus hernia, and sometimes in pregnancy. Heartburn in healthy sujbects can be due to hypersensitivity to some foods.

Nausea, the reflectory act associated with irritation of the vagus nerve,k is an indefinite feeling of cickness and sensation of compression in the epigastrium. Nausea is often attended by pallidness of the skin, general weakness, giddiness, sweating, salivation, fall in the arterial pressure, cold in the limbs, and ometimes semisyncopal state. Nausea often (but not necessarily) precedes vomiting. The mechanism of nausea is not know. Its frequent association with vomiting suggests that it might be the early sign of stimulation of the vomiting centre. The leading role in the development of uausea is given to the nervous system and also the tone of the stomach, the duodenum, and the small intestine. Nausea may develop without any connection with diseases of the stomach, e.g. in toxaemia of pregnancy, renal failure, deranged cerebral circulation, and sometimes in healthy people in the presence of foul odour (or in remembrance of something unpleasant). Some diseases of the stomach are attended by nausea, e.g. acute and chronic gastritis or cancer of the stomach are attended by nausea, e.g. acute and chronic gastritis or cancer of the stomach. Nausea associated with gastric pathology usually occurs after meals, especially after taking some pungent food. Nausea often develops in secretory insufficiency of the stomach.

Vomitting (emesis) occurs due to stimulation of the vomitting centre. This is a complicated reflex through the oesophagus, larynx and the mouth (sometimes through the nose as well).Vomitting may be caused by ingestion of spoiled food, by seasickness, or irritation arising inside the body (diseases of the gastro-intestinal tract, liver, kidneys, etc.). In most cases vomiting is preceded by nausea and sometimes hypersalivation. Factors causing the vomitting reflex are quite varied. This can be explained by the numerous connection that exists between the vomiting center (located in the medulla oblongata, in the inferior part of the floor of the 4th ventricle) and all bodily systems. Depending on a particular causative factor, the following can be differentiated: (1) nervous (Central) vomiting; (2) vomiting of visceral aetiology (peripheral or reflex); (3) haematogenic and toxic vomiting.

Vomiting is an important symptom of many diseases of the stomach but it can be regarded as the symptom of a particular disease only in the presence of other signs characteristic of this disease. Vomiting of gastric aetiology is caused by stimulation of receptors in gastric mucosa by inflammatory processes (acute or chronic gastritis), in ingestion of strong acids or alkalis, or food acting on the gastric receptors by chemical (spoiled food) or physical (overeating or excessively cold food) routes. Vomiting can also be caused by difficult evacuation of the stomach due to spasms or stenosed pylorus.

If the patient complains of vomiting, the physician should inquire about the time when the vomiting occurred, possible connections with meals, association with pain, the amount and character of the vomitted material. Morning vomiting (on a fasting stomach) with expulsion of much mucus is characteristic of chronic gastritis, especially in alcoholics. Hyperacid vomiting in the morning indicates nocturnal hypersecretion of the stomach. Vomiting occurring 10-15 minutes after meals suggests ulcer or cancer of the cardial part of the stomach, or acute gastritis. If vomitting occurs 2-3 hours after meals (during intense digestion) it may indicate ulcer or cancer of the stomach body. In the presence of ulcer of the pylorus or duodenum, vomiting occurs 4–6 hours after meals. Expulsion of food taken a day or two before is characteristic of pyloric stenosis. Patients with peptic ulcer often vomit at the height of pain thus removing it, which is typical of the disease. The odour of the vomit is usually acid, but it can often be fetid (puterefactive preocesses in the stomach); the odour may be even faecal in the presence of a faecal fistula between the stomach and the transverse colon).

The vomited material may have acid reaction (due to the presence of hydrochloric acid, in hyperchlorhydria), neutral (in achylia), or alkaline (in the presence of ammonia compounds, in pyloric stenosis, hypofunction of renal functin, and also in regurgitation of the duodenal contents into the stomach). Vomitus may contain materials of great diagnostic importance, e.g. blood, mucus in chronic gastritis),ample bile (narrowing of the duodenum, gastric achylia), and faecal matter. Vomiting may attend acute gastritis, exacerbation of chronic gastritis, gastric neurosis, peptic ulcer spasm and organic stenosis of the pylorus, and cancer of the stomach.

Pain is the leading symptom in diseases of the stomach, Epigastric pain is not obligatory connected with disease of the stomach. It should be remembered that the epigastrium is the "site of encounter" of all kinds of pain. Epigastric pain may be due to diseases of the liver, pancreas, and due to hernia of the linea alba. Epigastric pain may develop in diseases of other abdominal organs (sometimes of organ located outside the abdomen) by the viscero-visceral reflex (acute appendicitis, myocardial infarction, affection of the diaphragmatic pleura, etc.) In order to locate correctly the source of pain, the physician should ask the patient (1) to show exactly the site of pain (2) to characterize the pain which may be periodical or paroxysmal (at certain time of the day); permanent or seasonal (in spring or autumn); (3) to describe the connection (if any) between pain and meals, the quality of food and its consistency; (4) to indicate possible readiation of pain (into the back, shoulder blade, behind the sternum, left hypochondrium); (5) to describe conditions under which pain lessens (after vomiting after taking food or baking soda, after applying hot-water bottle or taking spasmolytics); (6) to describe possible connections between pain and physical strain (weight lifting, traffic jolting, etc.), or strong emotions. Intensity and character of pain are also imprtant diagnostically. The pain may be dull, stabbing, cutting etc. Pain is hollow organs with smooth muscles (e.g. stomach) is provoked by spasms (spastic pain), distension of the organ (Distensional pain), and by its motor dysfunction.

Paroxysmal, pepriodical epigastric pain is due to the spasm of the pyloric muscles. It arises under the influence of strong impulses arriving from the vagus nerve centre in cerebral cortex dysfunction. The spasm of the pylorus is stimulated by the hyperacidity of gastric juice due to hyperstimulation of the vagus. Depending on the time of paroxysmal pain (after meals), it may be early (occurring 30-40 min after meals),late (90-120 min after meals), nocturnal, and hunger pain (which is abated after taking food). If pain occurs after meals stimulating secretion of gasatric juice (bitter, pungent, spicy or smoked foods), this indicates the leading role of of hypersecretion in its aetiology. The pain then localizes in the epigastrium, radiates to the back, and is rather intense; it is abated after vomiting and taking alkali or foods that decrease acidity of gastric juice, and also after taking antispastic preparations and applying hot-water bottle (which removes spasms).

A seasonal character of pain, i.e. development of periodic pain during spring and autumn, is characteristic of peptic ulcer, especially if the process is localized in the peripyloric region. Permanent boring pain is usually caused by stimulation of the nerve elements in the mucous and submucous layer of the stomach; the pain is usually intensified after meals and is characteristic of exacerbation of chronic gastritis or cancer of the stomach.

Perigastritis (chronic inflammation of the peritoneum overlying the stomach and its adhesion to the neighbouring organs) is mainifested by pain developing immediately after taking much food (irrespective of its quality). The full stomach distends to stimulate nerve fibres in the adhesions. In the presence of perigastritis and adhesions between the stomach and the adjacent organs, pain may Be caused by any physical strain and when the patient changes his posture.

Gastric haemorrhage is a very important symptom. It can be manifested by vomiting of blood (haematemesis) or by black tarry stools (melaena). Gastric haemorrhage is usually manifested by the presence of blood in the vomitus. The colour of the vomitus depends on the time during which the blood is present in the stomach. If the blood was in the stomach for a long time, the blood reacts with hydrochloric acid of the gastric juice to form haematin hydrochloride. The vomitus looks like coffee grounds. If haemorrhage is profuse (damage to a large vessel) the vomitus contain much scarlet (unaltered) blood. Haematemesis occurs in peptic ulcer, cancer, and pylyps, in erosive gastritis, rarely in sarcoma, tuberculosis and syphilis of the stomach, and syphilis of the stomach, and in varicosity of the oesophageal veins. Tarry stools are not an obligatory signs of gastric haemorrhage.

When collecting **anamnesis**, the patient should be asked about his nutrition. It is important to establish if meals are regular because taking food at random is an important factor in the aetiology of gastric diseases. Food quality is an important as its amount taken during one meal. Masticationof food matters as well. Conditions of rest and work, and possible occupational hazards should be established. Abuse of alcohol and smoking are important factors in the aetiologyof gastric diseases. It is very important to find out if the patient's condition has undergone some changes during recent time (e.g. loss of weight, anaemia, blood vomiting, or tarry stools. Gastro-intestinal diseases of the past, surgical interention on the abdominal orgns, long medication with preparations irritating the stomach mucosa (acetylasalycilic acid, sodium salycilate, steroid hormones, potassium chloride, etc.) are also very important.

PHYSICAL EXAMINATION

Inspection

During general inspection of the patient the physician may assess poor nutrition of the patient (cachexia) which is characteristric of stomach cancer and untreated benign pyloric stenosis. Pale skin is observed after gastric haemorrhage. Patients with uncomplicated peptic ulcer look practically healthy.

Next stage is inspection of the mouth. The absence of many teeth accounts for inadequate disintegration and mastication of food in the mouth, while the presence of cariouos teeth favours penetrationof microbial flora into the stomach. The tongue is not the "mirror of the stomach" as it was formerly believed. Nevertheless in some diseases its appearance is informative: clean and moist tongue is characterstics of uncomplicated peptic ulcer, while thetongue coated with a foul smelling white-grey material is characteristic of acute gasatritis; a dry tongue indicates a severe abdominal pathology or acute pancreatitis; a tongue with atrophied papillae suggests cancer of the stomach, atrophic gastritis with pronounced gastric secretory hypofunction, or vitamin B deficiency.

Inspection of the abdomen may give information about the contours and peristalsis of the stomach if the patient is cachectic. In pathological cases (pyloric stenosis), peristalsis can be easily seen (ridges raising the abdominal wall). If the physician rubs or taps on the epigastric region peristalsis becomes more distinct. Sometimes, in neglected cases, the abdominal wall cn be protruded (tumour).

Palpation

The stomach should be palpated in both the vertical and horizonta position of the patient because the lesser curvature of the stomach anc its high standing tumours and impalpable in the lying position. Firs palpation should be superficial and tentative. Its aim is to establisl tenderness of the epigastrium, irritation of the periotoneum (Schetkin Blumberg symptom), divarication of the abdominal muscles, the presenc of hernia of the linea alba, tension in the abdominal wall in the regio: of the stomach, and the presence of muscular defence (defens musculaire). Deep palpation of the stomach should be carried ou according to Obraztsov and Strazhesko. The examiner pulls up th skin on the abdomen and presses carefully the anterior wall of th abdomen to penetrate the depth until the examining fingers reach th posterior wall. When pressed against the posterior wall of the abdomen the stomach slips from under the examining fingers. The shape of th stomach and the size of the examined part can thus be assessed. Th greater curvature and the pylorus can best of all be examined by thi: method.

The greater curvature can be examined by deep sliding palpatior in 50-60 per cent and the pylorous in 20-25 per cent of healthy subjects; the lesser curvature can be palpated in gastroptosis. The greater curvature is found to either side of the median lin2, 2-3 cm above the navel. It appears to palpating fingers as a ridge on the back bone and by its sides. In cases with gastroptosis, the greater curvature can descend below the naval. Correctness of determination can be confirmed if the position of the ridge coincides with that of the lower border of the stomach as determined by other techniques (by percussion, by the splashing soundor stethacoustic palpation). The pylorus is located in the triangle formed by the lower edge of the liver to the right of the median line, by the median line of the body, and the transverse line drawn 3-4 cm above the navel, in the region of the right rectus abdominis muscle. Since the position of the pylorus is oblique (upwards to the right) the palpating movements should be perpendicular to this direction, i.e. from left downwards to the right. The pylorus is identified by palpation as a band (tense or relaxed). When the pylorus is manipulated by the fingers, a foft rumbling sound can be heard. When contracted spastically (pylorospasm) the pylorus remans firm for a long time. Sometimes the pylorus is mistaken for cancer infiltration.

Palpation of the stomach can reveal tumours of the pylorus, of the greater curvature, and of the anterior wall. Tumours of the lesser curvature can be diagnosed with the patient in theupright position. Tumours of the cardial part of the stomach are inaccessible to palpaton. Exact information of their location gives X-ray examination.

Percussion

Percussion is used to determine the inferior border of the stomach. Provided professional skill is high, the inferior border of the stomach can be outlined by light percussion by differentiating between gastric and intestinal tympany.

If the patient cannot eat the full meal (The capacity of the stomach gradually decreases), it is necessary to determine the Trabue's space, which can be markdely decreased. The presence of these two symptooms requires an X-ray examination to exclude cancer of the stomach (scirrhus). Short strokes of the hammer or the flexed fingers on the epigastrium (Mendel sing) are used to determine involvement of the parietal periotoneum: pain indicates affection of the periotenum.

Splashing sound (succession) can be heart if the patient is lying on his back, while the examiner pushes the anterior wall of the peritoneum with four flexed fingers of the apt hand. The outer hand of the physician should fix the muscles of the abdominal prelum against the sternal edge. The thrust of the hand is transmitted through the stomach wall to the liquid and air contained inside it to cause a readily audible splashing sound which is inaudible outside the inferior border of the stomach. This technique for outlining the inferior border of the stomach is effective in casaes where the stomach border formed by the greater curvature is at the normal level or lowered. Succession gives information about the evaculatory function of the stomach: the splashing sounds in healthy subjects can only be heart after meals. Splashign sounds heart 7–8 hours after meals suggest evacuatory dysfunction of the stomach (mostly in pyloric stenosis) or its pronounced hypersecretion (gastrosuccorrhoea). Splashing sounds heart to the right of the median line of the abdomen indicate dilation of the prepyloric part of the stomach (Vasilenko's symptom)..

Auscultation

Ausculation of the stomach is practically non-informative. It is only helpful when used together with palpation of the stomach to outline its inferior border. Stethacoustic palpation is performed as follows: a stethoscope is placed beneath the left costal arch, below the Traube's space. The examiner rubs the abdominal wall overlying the stomach by the finger and gradually moves the finger away from the stethoscope bell. As long as the finger rubs the skin overlying the stomach, the physicain hears the friction, but when the finger moves outside the stomach borders, the sound disappears. This method is very simple but the findings are sometimes inaccurate.

LABORATORY AND INSTRUMENTAL METHODS

Examination of the Secretory Function

Study of gastric secretion is an indispensable part of comples diagnosis of the gastric mucosa function. The most reliable data on the gastric secretion can be obtained by studying gastric juice.

A thick tube was used for many years to study gastric secretion. The main disadvantage of this method is that a mixture of gastric juice and test meal is extracted and it is thus impossible to obtain reliable qualitative and quantitative characteristics of gastric secretion. This method can therefore be considered obsolete.

SPECIAL PATHOLOGY

Gastric peptic ulcer, and cancer are most common diseases of the stomach.

GASTRITIS

Gastritis is inflammation of the gastric mucous membrane. Acute and chornic gastritis are differentiated.

Acute Gastritis

Acute gastritis is a very common disease. Catarrhal, corrosive and acute phlegmonous gastritis are distinguished.

Aetiology and pathogenesis

The alimentary factor is the leading one among the other causes of gastritis. Ingestion of fatty, coarse, poorly assimilated or decomposed food, of very cold or hot meals or alcoholic drinks provokes gastritis. The gastric mucosa may also be irritated by prolonged taking of some medicines. An allertic reaction to some foods (fish, eggs) can sometimes also cause gastritis. Various infectious diseases, such as influenza, measles, or scarlet fever can be attended by acute gastritis. It is the leading symptom in food poisoning.

Clinical picture

Symptomatology of acute gastritis is quite varied. It may proceed asympatomatically or be manifested by severe local and general symptoms. Inflammations begins 2-3 hours after the irritating agent has been ingested while the clinical picture develops in 6-8 hours. The symptoms of acute gastritis are loss of appetite, unpleasant taste in the mouth, nausea and vomiting (first with food remains and later with bile).The patient complains of pressure, bulging,k and pain in the epigastrium. The rise of temperature is often preceded by chills. The patient is pale, his pulse is acelerated, the tongue coated, and breath is foul. Palpation of the epigastric region is painful and stimulates nausea. The amount of excreted urine decreases. In persistent vomiting the haemoglobin and erythrocyte counts increase while the level of chlorides decreases.

The onset of the disease is characterized by hypersecretion and hyperchlorhydria which is followed by inhibition of gastric secretion. The motor (evacuatory) function of the stomach sharply decreases, while the absorptive power of the gastric mucosa inceases. Gastroscopy reveals hyperaemic mucosa coated by a thick layer of glassy mucus; haemorrhage and erosion sometimes occur. Morphological reconstruction of the gastric mucosa complete by the end of the second week. The patient revovers clinically in 3 or 4 days. X-ray examination fails to reveal any changes in the relief of the gastric mucosa.

Course

The disease ends by recovery in most cases, but acute process can convert into a chronic one in the absence of appropriate treatment and if exposure to the harmful factors (alcohol, overreating) continues.

The special forms of acute gastritis (corrosive and phlegmonous) are called fulminan tgastritis. Corrosive acute gastritis develops from irritation of the gastric mucosa by acids or alkalis, and acute phlegmonous gstritis is a rare disease characterized by purulent inflammation of a part of the entire stomach wall. The clinical course of both forms is grave. Perforation of the stomach is possible, Prognosis is unfavourable in most cases.

Treatment

During the first 2 or 3 days the patient is given only nutritive liquid, and the diet is then gradually enriched by the end of the second week. Bed rest is indicated during the first days on the disease; antacid and astringent preparations are given. The stomach should be lavaged as soon as possible in corrosive gastritis. Antibiotics should be given in phlegmonous gastritis.

Chronic Gastritis

Chronic gastritis is a frrequently occurring disease; it affects mostly men.

Aetiology and pathogenesis

Chronic gastritis is closely connected aetiologically with acute gastritis. All exogenous aetiological factors responsible for acute gastritis are important for the development of chronic gastritis. Among important endogenous factors causing chronic gastritis are reflex effects on the stomach from pathologically changed organs (gall bladder, intestine, pancreas), upset hormone system (affection of the thyroid, pituitary, adrenal glands), chronic infections (tuberculosis, malaria, syphilis), chronic septic foci (tonsillitis, carious teeth). Chronic gastritis may develop from disordered metabolism (diabetes, obesity, gout, renal failure). Finaly, chronic gastritis may be secondary to diseases accounting for hypoxia of tissue, the gastric mucosa included (chronic circulatory insufficiency,k pulmonary heart, uraemia).

Chronic gastritis is a polypathogenic disease. The leading role in its development undoubtedly belongs to disorders in the nervous, humoral, and hormonal mechanisms regulating the digesting function. But the direct action of irritants on the gastric mucosa cannot be disregarded either. Autoallergic processes are important for maintaining the disease.

Pathologicl anatomy

Differentiated are superficial affections of the mucosa, affection of the glands without their atrophy, atrophy of the gastric mucosa of various degree, and combination of atrophy with hyperplasia.

Classification of chronic gastritis

There is no universally accepted classification of chronic gastritis. S. Ryss proposed a classification based on four principles: aetiological (exogenous and endogenous), morrphological, functional (chronic gastritis with preserved secretion and with secretory insufficiency of various degree, up to achlorhydria), and clinical (compensated chronic gastritis or remission phase; decompensated gastritis or exacerbation phase). Moreover, gastritis can be regarded as an independent disease and as a disease secondary to other pathologies (peptic ulcer, chronic colitis, etc.).

Clinical picture

The signs of chronic gastitis ar difficult to describe because the course and symptomatology of the disease are quite variable. Some patients do not complain of anything during remissions; the disease may also develop for a long time without any manifestations and it is therefore difficult to establish the time of its onset.

The main syndrome of chronic gastritis is gastric dyspepsia. It may combine with intestinal dyspepsia characterized by meteorism, rumbling sounds in the abdomen, constipation, and diarrhoea. The patient with chronic gastritis can also feedl pressure and distention in the epigastrium, and sometimes pain. These symptoms are connected with distention of the stomach by ingested food and the attending pathological sensitivity of mucosal interoceptors. Pain is dull and boring, but sometimes becomes severe. The general condition of patients with chronic gastritis varies. Some patients do not lose weight, remain active, while others lose weight, become flaccid and slow, their appetite is poor. A pronounced decrease ini gastric secretion may be attended by diarrhoea which causes even greater wastin gand impairs absorption of proteins, vitamins, and iron. Anaemia develops along with signs of polyhypovitaminosis and albumin deficiency.

Examination of the abdomen sometimes reveals inflation. Palpation of the epigastrium is in most cases painless. The acid secretion may remain normal or it may decrease. Free hydrochloric acid may be absent from the gastric juice (Achlorhydria). In neglected cases secretion of pepsin (achylia) is upset as well.

Roentgenorgraphy is but of littler use in the diagnosis of chronic gastritis. Gastroscopy can give valuable diagnostic information, especially if it is combined with sighting biopsy. Aspiration biopsy an dexfoliative cytology are also important for the study of patients with chronic gastritis.

Chronic gastritis should be differentiated from peptic ulcer and gastric neurosis ("irritated stomach") in the presence of hyperchlorhydria and functional achylia.

Course

The disease usually slowly progresses but the appropriate treatment improves the patient's condition. Complete anatomical restorattio of the gastric mucosa and normalizaton of the secretory function occur rarely. Chronic gastritis with secretory insufficiency is considered as a precancer condition.

Treatment

Therapy of chronic gastritis should be combined, differential, lengthy, and planned. Diet is especially important. When prescribing an appropriate diet, the stage of the disease. and also the secretory background of the stomach should be taken into consideration.

Substitution therapy (preparations of gastric juice) is given in anacid gastritis. In spastic and pain syndromes, spapsmolytics are given and thermotherapy applied. Health-resort and sanatorium therapy are also indicated in combined treatment.

Prophylaxis

Rational nutrition and observation of hygienic requirements are requisite in prevention of chronic gastritis. Fighting against smoking and drinking alcohol, sanation of chronic inflammatory foci (carious teeth, tonsillitis)) should also be included into the list of preventive measures. Patients with chronic gastritis should be regularly observed in out-patient conditions.

Peptic Ulcer Disease (Gastric and Duodenal Ulcer)

Peptic ulcer is a general chronic and relapsing disease characterized by seasonal exacerbations with ulceration of the stomach wall or the duodenum.

Aetiology and pthogenesis

The aetiology and pathogenesis of the disease is still unknown despite and intense clinical and experimental research.

The main affection, the lession in the wall of the stomach or duodenum, is caused by the action of gastric juice. But in normal conditions the gastric or duodenal mucosa is resistant to the digestive effect of the juice due to the presence of complex protective mechanisms. Therefore, in order to cause self-digestion, certain factors should be involved which decrease resistance of the muocsa to the digestive effect of the gastric juice or increase its digestive properties; or both these factors should be involved. Many theories have been proposed to explain these phenomena but none of them fully explains the causes of peptic ulcer disease.

From the findigns available at the present time, the following main factors in the pathogenesis and aetiology of the disease have been established: (1) disordered neurochromal mechanisms regulating digestion; (2) disorders in the local digestive mechanisms; (3) structural changes in the gastric and duodenal mucosa.

The predisposing factors are heredity and environmental factors, among which nutrition is the leading one. Irregular nutrition, with prevalence of easily assimilable carbohydrates in the diet, excess ingestion of poorly assimilated and long digested foods cause hypersecretion of the stomach. In the presence of the main factors, the predisposingi factors cause ulceration with time. Alcohol and nicotine have also an adverse effect on the gastric mucosa.

The central role in the aetiology and pathogenesis of peptic ulcer belongs to disorders in the nervous system which can arise in the central and vegeative systems under the action of various effects (negative comtions, physical and mental metnal overstarin, viscero-visceral reflexes, etc).

Current research shows the imprtant role of endocrine dysfunction in the development of peptic ulcer (dysfucntion of the pituitary or the adrenal glands). Disorders in local mechanisms (the acid-peptic factor, gastric hormones, mucous barrier, regeneration of mucosa, blood circulation in the stomach wall and the duodenum, morphological reconstruction of the mucosa, changes in the motor function, condition of local secretory depressor mechanisms) result from disordered neurohormonal regulation of the gastroduodenal system.

It can thus be concluded that the cause of peptic ulcer are varied, while its pathogenesis is complex and uncertain. Nervous genesis may probably prevail in some cases, while humoral or neurohumoral in others; disorders in the local mechanisms of gastric digestion may also be the leading factor in the aetiology and pthogenesis of peptic ulcer

Pathological anatomy

Margins of a chronic ulcer are consolidated. Ulcers with especially hard elevated edges are called callous. Inflammatory infiltrations are usually formed round the ulcer. A bleeding vessel can be found on its floor. An ulcer can destroy the wall of the stomach (perforating ulcer) or it can penetrate the adjacent organs.

Classification

There is no universally accepted classification of peptic ulcer disease. Location of an ulcer determines to a certain degree the clinical couse of pathology. Ulcers mostly develop on the lesser curvature. The cardia is affected less frequently. The usual site of ulcer in the duodenum is the bulb. Pyloric, postbulbar, and extrabulbar ulcers are also differentiated. Age and sex are important for the clinic of the disease. Hence juvenile ulcers, ulcers of the elderly and old patients, and also peptic ulcer disease of women are differentiated.

Clinical picture

Symptoms of peptic ulcer disease vary and depend on the age, sex, the general condition of the patient, the duration of the disease, frequency of exacerbations, location, kind of lesion, and the presence of complications.

The leading symptom of peptic ulcer is pain, which may be periodic, seasonal, increasing in severity, intensifying or lessening after vomiting or taking meals, alkalis, thermal procedures, or cholinolytic preparations.

Early pain is typical of gastric ulcer; late pain, nocturnal or hunger pain are characteristic of peripyloric and duodenal ulcer. Permanent pain is atypical and is usually due to complications (perivisceritis, penetration of the ulcer). Regular connection between pain in peptic ulcer and quantity and quality of food is obvious. Ample, bitter, sour, salty, spicy and coarse food always cause severe pain.

The seasonal character of pain (vernal and autumnal) is very typical of peptic ulcer disease and can positively be used to differentiate it from pain in other diseases. Exacerbations of pain are alternated with remissions even in the absence of treatment. The cyclic character of peptic ulcer disease is probably due to seasonal changes in general reactivity of the human body. Upset vitamin balance in spring may play a certain role in the course of the disease as well. Periodicity of pain may be not quite obvious at the early stage of the disease. Except hunger pain, which lessens after meals, pain attains its maximum severity at the height of digestion.

Although a lesion cannot always be located by pain, it has, however, been noticed that gastric ulcer is manifested by pain in the epigastrium above the navel, while in duodenal ulcer pain is felt in the epigastrium to the right of the median line; ulcer of the cardia is characterized by pain in the vicinity of the xiphoid process. Pain may radiate into the left breat nipple, behind the sternum, the left shoulder blade, and into the thoracic part of the spine.

The pathogenesis of pain in peptic ulcer is uncertain. It has been established that common stimuli causign pain when applied to the skin are ineffective when applied to the wall of the stomach or the intestine. Irritating factors for these organs are muscular strain, especially spasms, and elevation of the intravisceral pressure, which are caused by disorders of nervous regulation.

The acid factor acting on the motor function of the stomach and the duodenum, and also chanages in the mucosa of these organs are very important in the aetiology of pain in peptic ulcer disease.

Vomiting occurs in 70–75 per cent of patients. It arises without preliminary nausea, at the height of pain, and relieves it. The vomitus is acid to taste and smell. Secretion of the gastric juice in a fasting stomach is often attended by vomiting as well.

Heartburn is encountered in 60–85 per cent of patients. It occurs not only during exacerbations but can also precede exacerbation for several years; it can be periodic or seasonal. The mechanism of heartburn is associated with motor dysfunction of the oesophagus (in addition to the acid factor of the gatric contents, which was formerly believed to be decisive).

Eructation, regurgitation, and salivation are frequent symptoms. Appetite is often increased. Regular connection between meals and development of pain is the cause of a morbid fear of eating (cibophobia).

The intestinal symptoms of peptic ulcer disease are constipations, which are closely connected with the character of nutrition and bed-rest during exacerbations, and are mainly connected with reflex dyskinesia of the small and large intestine.

Wasting is characteristic of exacerbations. The skin and mucosa are pallid after haemorrhage. The tongue is usually clean. The configuration of the abdomen is normal. In the presence of pyloric stenosis peristaltic and antiperistaltic movemetns of the epigastrium can be seen. Brown pigmentation develops on the abdomen after prolonged application of warmth. During exacerbations, the epigastric region is tender to surface palpation; if the peritoneum is involved (positive Mendel's test) the muscles are strained. Late splashing sound to the right of the median line (Vasilenko's symptom) indicates gastric evacuatory dysfunction or increased secretion between meals.

Gastric secretory function

If the ulcer is found in the stomach hydrochloric acid, pepsin, mucoprotein and albumin fractions of the gastric juice vary within normal limits. In duodenal ulcer all these indices significantly exceed normal values. Hypersecretion of the gastric juice is determined in this case by hyperpsensitivity of the vagus, intensified adrenal function, and increased quantity and hypersensitivity of the parietal cells. Study of the basal secretion of a fasting stomach is very important for the diagnosis of duodenal ulcer. In the prepsence of active basal secretion and specific complaints of the patient, he should be given the appropriate treatment despite negative results of X-ray examination.

The *motor funciton of the stomach and duodenum* is upset in peptic ulcer. This is the leading factor in the development of the main symptoms, e.g. pain, nausea, vomiting, and heartburn. Upset motor function increases the tone of the stomach and the duodenum, intensifies their peristalsis, and disturbs periodicity of their activity. The observed changes in the motor function are not specific for peptic ulcer disease and cannot therefore be regarded as its diagnostic signs. But a certain regularity of these changes can indicate the stage of the disease and be a criterion for the objective judgement about the efficacy of treatment.

Latent haemorrhage is almost always revealed on examination of faeces during exacerbation of peptic ulcer.

X-ray examination

A direct proof of peptic ulcer is a niche which is found in 75-80 per cent of patients. The ulcer is usually located on the lesser curvature. It has the regular J shape, the barium depot extending beyond the normal contour of the stomach. In duodenal ulcer, the niche can be found inside the bulb or outside it (extrabulbar ulcer).

In the absence of direct X-ray signs (the absence of a niche), indirect symptoms become important. These are intensified peristalsis of the stomach, and the presence of a thick layer of secretion between the air and the contrast substance.

Gastroscopy

Gastroscopy is used to reveal gastric ulcer; if gastroscopy is repeated, it shows the cicatrization process. A fibroscope can be used not only to view the stoamach, but the duodenum as well. Ulcers that fail to be detected by X-rays can be revealed by gastroscopy, which is also important for differentiation between benign and malignant changes (sighting biopsy).

Course

Four stages are distinguished in the clinic of peptic ulcer: stage I ("prelude to ulcer", according to M. Konchalovsky) is characterized by a marked disorder in the activity of the vegetative nervous system and gastroduodenal dysfunction; stage II is attended by development of organic changes (gastroduodenitis); stage III is formation of an ulcer; and stage IV is development of post-ulcerous processes.

This classification of ulcer stages is only conventional but still useful because it may help the physical to establish early diagnosis. Remission periods last from several months to many years. Exacerbations continue for 4–6 weeks. Cicatrization of the ulcer is completed in 6–8 weeks. Remission may occur without treatment. Seasonal exacerbations are more characteristic of duodenal ulcer. Unless complicated by haemorrhage or perforation, peptic ulcer is never fatal.

Haemorrhage

This is the most frequent complication. It may be manifested by haematemesis (blood vomiting) and tarry faeces (melaena). Among other causes of gastric haemorrhage, peptic ulcer is accounted for 60-65 per cent. Gastric ulcers bleed more often than others. Vomiting may be absent in duodenal ulcer, and the first signs of haemorrhage are sudden weakness, giddiness and palpiptation (before the appearance of tarry stools). The patient's general condition depends on the length and intensity of bleeding.

Perforation

Predisposition to open perforation depends mainly on anatomical factors: the probability of perforation is especially high in location of the ulcer on the anterior wall of the duodenum. Perforation may occur in the absence of patient's complaints ("silent ulcer") or the appropriate anaemnesis. Signs of perforation are a sudden stabing pain, the reflex collase, acute abdomen, and progressive peritonitis (unless a timely surgical aid is given to the patient). The pain is felt beneath the xiphoid process or in the right hypochondrium. The abdominal wall is tense. The patient assumes a forced posture on his back; the tongue is dry and coated. The pulse is retarded, the temperature is subnormal. In most cases the diagnosis is undoubtful and a timely surgical intervention saves the patient's life.

Stenosis

Ulcers heal to leave scars. If the ulcer was in the pylorus, the cicatricial tissue may narrow the lumen and interfere with free passage of the gastric contents into the duodenum. First the narrowing is compensated for by hypertrophy of the gastric muscles, but later the stomach becomes distended, food stays inside it for a longer period, and fermentation and putrefaction occur in the stomach. Absorption of water is impaired (the impairment begins in the duodenum). This upsets the water-salt balance, causes general dehydration of the body, and decreases the chloride content in the blood and urine.

Patients complain of permanent pain (which intensifies by night), eructation with rotten egg wind, and profuse morning vomiting with food that was ingested several days ago. Constipation is alternated with diarrhoea, through irritation of the small intestine by fermented food discharged into it from the pylorus which is opened by intensified peristalsis. If stenosis is pronounced, the patient is cachectic. Examination of the epigastrium reveals peristaltic and antiperistaltic contractions of the stomach. Late splashing sound can be heard.

Treatment and prophylaxis

Anti-ulcer treatment implies combined and individual therpay. Non-complicated peptic ulcer is treated conservatively. Exacerbations should be treated in hospital where bed-rest, special diet, medicinal preparations (tranquilizers, cholinolytics, antacids) and thermal procedures are given. The diet should not stimulate the secretory function of the stomach. It should decrease the motor function of the gastroduodenal system and have buffer properties. The diet should also be sparing with respect to its chemical and mechanical effect on the stomach.

Patients in remission should be regularly observed in out-patient conditions. Smoking and alcohol are prohibited in peptic ulcer disease. Attending diseases should also be teated and sanation of the mouth carried out. Sanatorium and health-resort therapy is indicated. Employment of patients should be sparing. Persons with "irritated" stomach (functional stage of peptic ulcer) should be regularly observed in out-patient conditions.

Prophylactic measures should be aimed at prevention of the disease, its relapses or complications. Current knowledge of the developmental mechanisms of peptic ulcer suggests that population should be given information on rational nutrition, labour, and rest. The adverse effect of alcohol and smoking should be especially emphasized.

INADEQUATE DIGESTION SYNDROME

This is a symptom complex characterized by digestive disorders in the gastrointestinal tract. Disorders can be associated with upset cavital digestion, i.e. in the stomach and the intestine (dyspepsia), and parietal digestion. Mixed forms also occur.

Aetiology and pathogenesis

Dyspepsia

This occurs due to non-compensated secretory insufficiency of the to mach, exocrine dysfunction of the pancreas, upset secretion of bile, disordered passage of chyme through the gastrointestinal tract (stasis, congestion due to setnosis and compression of the intestine, or accelerated passage due to intense peristalsis). Intestinal infection, sysbacteriosis, and alimentary disorders (overeating, diet rich in protein, fats, or carbohydrates, intake of large amounts of fermented drinks) are also important. *Dyspepsia* may be functional, but commonly it is due to a digestive tract disease.

Pathogenesis: incomplete breakdown of food particles, active propagation of bacteria flora in the intestine with its invasion of the proximal parts of the small intestine, dysbacteriosis, abnormally high activity of bacteria in the enzymatic decomposition of food with formation of toxins (ammonia, indole, low molecularr fatty acids, etc.) that irritate the intestinal mucosa, intensify peristalsis, and cause symptoms of toxaemia due to their penetration into the blood.

Clinical picture

Gastric dyspepsia

This condition occurs in *achlorhydria* and *achylia*, long-standing decompensated pyloric stenosis, atrophic gastritis, and cancer of the stoamch. The disease is characterized by the feeling of discomfort, pressure or distension in the epigastrium after meals, frequent eruct-ation, regurgitation (often with acid or fetid odour), unpleasant taste in the mouth, nausea, and poor appetite. Achylous diarrhoea and meterosim are not infrequent. Examination of the gastric juice reveals achlorhydria or achylia (both hydrochloric acid and pepsin are absent from the gastric juice).

Intestinal dyspepsia occurs in the presence of exocrine dysfunction of the pancreas, chronic inflammatory diseases of the small intestine, and some other conditions. It presents as inflation and rumbling in the abdomen (borborygmus), intensive passage of flatus, diarrhoea withputrefactive or acid smell and (in rare cases) constipation. Corprologic findings: steatorrhoea, amylorhoea. X-ray findings: acclerated passage of the barium meal through the small intestine. Studies of the secretory function of the pancreas, aspiration biopsy, determination fo enterokinase and alkaline phosphatase in the intestinal juice, and othe rtests help verify the cause of intestinal dyspepsia.

Tests for hyperglycaemia with an oral starch load and radionuclide studies with glycerol trioleate, sunflower-seed or olive oil are used to estimate the degree of the cavital digestion derangement. The study of intestinal microflora is also important.

Treatment

This, in the first instance, is aimed at eradication of the underlying disease. Symptomatic therapy includes special diets and medicationfor diarrhoea; enzyme preparations (pancreatin, abdomin, festal), astringents, and carbolen are also administered.

Inadequate parietal digetion can be seen in congenital secretory insufficiency of the intestinal wall (disaccharidase deficit enteropathy) and in chronic diseases of the small intstine attended with dystrophic, inflammatory, and sclerotic changes in its mucosa, upset structure of the villi and microvilli, and their decreasing number, deranged intestinal peristalsis (enteritis, sprue, intestinal lipodystrophy, exudative enteropathy, etc.).

The symptoms, course, and clinical picture are the same as in intestinal dyspepsia and in the malabsorption syndrome.

The diagnosis is established on the basis of determination of enzymes (amylase, lipase) during their desorption in homogenous preparations of the small-intestine mucosa specimens taken by aspiration biopsy. The glycaemic curve drawn after oral intaks of disaccharides and monosaccharides helps differentiate the syndrome of inadequate parietal digestion (flat curve after the intake of maltose, saccharose, lactose; normla curve after the glucose and galactose intake) from lesions of the small intestine that are attended with malabsorption of the products of food decomposition in the small intestine. After an intake of polysaccharides (starch) this syndrome may be differentiated from inadequate cavital digestion. Aspiration biopsy reveals atrophy of the intestinal mucosa (indirect sign)

Treatment is aimed at the main disease. For methods of management of malabsorption syndrome see below. Symptomatic therapy includes enzyme preparations (abomin, festal) and astringents (tannalbin, albin). The drugs are given per os.

MALABSORPTION SYNDROME

This symptoms complex is due to upset absorption in the small intestine. The syndrome often develops in combination with the syndrome of inadequate digestion. Primary and secondary malabsorption symdromes are distinguished.

Malabsorption is probably explained by congenital disorders in the fine structure of the intestinal mucosa and genetically determined intestinal enzymopathy. The onset of the secondary malabsorption syndrome is due to acquired structural changes in the intestinal mucosa evoked by acute and chronic enteritis, sprue, intestinal lipodystrophy, exudative enteropathy, lesions of the small intestine in amyloidosis, systemic scleroderma, and other disease conditions attended with digestive disorders. The upset intestinal digestion and accelerated passage of chyme through the intestine are decisive in acute and subacute diseases. In chronic conditions the decisive factors are dystrophy and atrophic fibrous chanages in the mucosa of the small intestine, shortening and levelling of the villi and crypts, significant reductionof the microvilli number, fibrous tissue formation in the intestinal wall with impairment of blood and lymph circulation, and disorderd parietal digestion. All these changes limit absorption of the products of hydrolysis of proteins, fats, and carbohydrates, and of mineral salts and vitamins in the intestinal wall.

Clinical pictures. Course. Gradual wasting, symptoms of metabolic disorders of all types (protein, fat, vitamin, water-salt), dystrophic changes in the internal organs with their subsequent dysfunction, and also constant steatorrhoea, creatorrhoea, and amylorrhoea are characteristic. Hypoproteinaemia develops (mostly at the expense of reduction of the serum albumin level): hypocholesterolaemia, hypocalcaemia, and moderate hypoglycaemia occur. Hypoproteinaemic oedema develops in the presence of hypoproteinaemia below 40-50 g/l. The characteristic symptoms of poly-hypovitaminosis are osteoporosis, anaemia (hypochromic anaemia in predominant malabsorption of iron, and hyperchromic anaemia in upset absorption of vitamin B_{12}), trophic changes in the skin, nails, progressive atrophy of the muscles, signs of polyglandular insufficiency, weakness, and (in severe cases) acidosis and cachexia.

Diagnosis

Laboratory examinations determine hypoproteinaemia, hypocholesterolaemia, hypoglycaemia, and other disorders due to malabsorption.

Corprologic studies reveal increased content of undigested food in the faeces and also increased excretion of the products of enzymic decomposition of food. Enterobiopsy reveals atrophic changes in the mucosa of the proximal parts of the small intestine. Since the walls of the small intestine absorb great amounts of various substances, different methods are used to study their absorption. These are tests with carotine, folic acid, galactose, D-xylose absorption test, etc.

Caseine, albumin, oleic acid, methionine, glycine, vitamin B12, folic acid, and other substances labeled with radioactive isotopes have recently been used. The method is based on the determination of concentration of labelled substances and the time of their appearance in the blood, their excretion with the urine or faeces, and assessment of residual radioactivity of faecal masses that is indicative of the amount of unabsorbed substances. Determination of the absorbed nutrients is based on the study of the chemical composition of food and stools during a certain period of time.

The course of the disease depends on the underlying condition and prospects for its cure. The prognosis is unfavourable in severe cases.

Treatment

The underlying disease should be treated. Symptomatic treatment: parenteral nutritional, administration of vitamins, plasma, protein hydorlyzates, glucose, nutrient enema, correction of electrolyte metabolism.

Prophylaxis includes timely treatment of disease that usually concur with the malabsorption syndrome.

SPECIAL PATHOLOGY

Most common diseases of the intestine are dyskinesia, inflammatory affectiosn (enteritis, colitis, enterocolitis) and tumours (mostly cancer of the large intestine).

According to their clinical course, inflammation of the small intestine (enteritis) and of the large intestine (colitis) may be acute or chronic.

Acute Enterocolitis

Acute inflammation of the small and large intestine usually combines with affection of the gastric mucosa and arises after ingestion of spoiled food infected with microorganisms or after ingestion of a large amount of hardly digestable or incompatible foods (gastroenterocolitis).

Clinical picture

The clinic of acute enterocolitis varies from mild illness to fatal outcomes. The onset of the disease is sudden (3–4 hours) following ingestion of inadequate food). Its first symptom is dyspepsia (diarrhoea). The body temperature is subfebrile or higher. The tongue is dry and the abdomen distended, tenderness is diffuse. Acute symtoms subside in 8–12 hours and the patient's condition improves in few days. Collapse may occur in severe cases due to poisoning.

Treatment

The stomach should be lavaged and purgative salts given. Sulpha drugs are given with a special diet; subcutaneous injections of sodium chloride are useful in marked dehydration.

Prophylaxis

This consists in adequate hygiene of nutrition, thorough inspection of cooking and foods storage conditions, especially during hot seasons.

Chronic Enteritis

Aetiology and pathogenesis

Chronic enteritis arises due to various causes. These are (1) infection: typhoid fever, dysentery, salmonellosis, etc.; (2) acute enteritis (a aforerunner); (3) dysbacterosis: upset microbial equilibrium in the intestine; (4) alimentary factor: irregular meals, ingestion of cold food, chronic overeating of poorly digestable foods; (5) radioactive exposure; (6) alcohol abuse; (7) allergic factors; (8) congenital enzymopathy; deficient quantity of enzymes responsible for absorption of foods (gluten and lactase deficiency); (9) endocrine factors (diarrhoea in thyrotoxicosis); (10) diseases of other alimentary organs (stomach, hepatobiliary system, pancreas). For example, in the presence of achlorhydria, insufficiently digested food enters the small intestine to irritate its mucosa and to provide inflammation.

Pathological anatomy

Mucosa of the small intestine is oedematous and hyperaemic. Haemorrhage and ulceration are possible. In grave cases, inflammation may involve all layers of the intestinal wall to cause its perforation.

Clinical picture

The patient usually complains of pain in the umbilical region and distensionof the abdomen. Stools are not formed; constipations are alternated with diarrhoea. Nutriton is impaired, the skin is pallid. Signs of polyhypovitaminosis are present: dry skin, brittle and laminated nails. Splashing and rumbling sounds are heard in the right iliac region. Stools contain mucus; microscopy of faeces reveals the presence of drops of neutral fat and muscle fibres. Specific X-ray signs are hypotonia, the presence of gas and liquid in the small intestine, and level relief or feather like pattern of its mucosa.

Course

Chronic enteriteis can be complicated by involvement of the pancreas, the liver, the large intestine, by development of hypochromic anaemia, and polyhypovitaminosis.

Treatment

Complex therapy is required. It is necessary to take into account the degree of the peptic disorder, complications if any, and the general conditions of the patient. In exacerbation of the disease, sulpha perparations, eubiotics, and enzyme preparations (abomin, pancreatin, pansinorm, festal, and others) are indicated.

Prophylaxis

This consists in eradication of possible causes of the disease and timely and thorough treatment of acute enterocolitis, chronic gastritis, diseases of the liver and the pancreas.

Chronic Colitis

Aetiology and pathogenesis

Causes of inflammatory affections of the large intestine are quite varied. Most frequent causes of chronic colitis are infections (dysentery, salmonellosis, tuberculosis, syphilis, etc.), parasites (helminths, protozoa, etc.) and toxic effects (poisoning with arsenic, phosphorus, mercury, etc.). Irregular nutrition, overeacting, and chronic constipations can account for development of colitis as well.

In the presence of motor hyperfunction of motor hyperfunction of the small intestine, the ingested food is nto processed sufficiently before it enters the large intestine, and it thus irritates its mucosa. Loong-standing kinetic disorders cause colitis. Persistent constipations can provoke chronic colitis. Mucosa of the large intestine has the excretory function. It releases microbes and their toxins, i.e toxic products that circulate in the body in cases with upset metabolism. These factors can become the cause of chronic colitis, e.g. renal dysfunction causes development of colitis. And finally, auto-infection (e.g. coli bacilli which become pathogenic under certain conditions) can stimulate the onset of colitis.

Classification of colites

The following colites are distinguished: I, infectious colites: (1) specific, and (2) non-specific; II, parasitary colites: (1) protozoal (amoebic, trichomonal, lambliogenic), (2) helminthic; III, toxic colites: (1) exogenous and (2)) endogenous; IV, alimentary colites; V, symptomatic or secondary colites; and VI, colites of mixed aetiology.

Pathological anatomy

The entire large intestine or its separate sections may be affected by inflammation. Catarhal, follicular, infiltrative, purulent, ulcerative, and gangrenous colites are differentiated from the standpoints of pathological anatomy.

Clinical picture

The patient with chronic colitis complains of local and general disorders. Local complaints include pain the lower abdomen or the iliac region, distension of the abdomen, tenesmus, constipation and diarrhoea. General complaints are irritability, deranged sleep, headache, and low moods. Appetite is decreased, nausea and sometimes vomiting occur. Objective examination shows that nutrition is adequate. The study of the abdominal cavity reveals pain by th ecourse of the large intestine and rumbling. Protozoa or helminths can be found in faeces. Stools may also contain traces of blood and mucus; dysbacteriosis is also possible. X-ray examination may reveal spasms, atonia of separate portions of the large intestine, and changes in the relief of the intestinal mucosa. Rectosimoidoscopy and colonoscopy are valuabale diagnostic techniques.

According to the clinical course of the disease, mild, medium gravity and grave chronic colites are distinguished. Mild forms of colitis have not pronounced symptoms; only occasional diarrhoea or constipations are observe; the general condition of the patient is not affected substantially.

Signs of the disease are pronounced in chronic colitis of medium gravity. Grave forms of the disease are marked by fever, headache, asthenia, disability, involvement of other organs, and complications (haemorrhage, perforation).

Treatment

A correct treatment is only possible if the cause of chronic colitis has been discovered. Changes in other organs of the digestive system and the presence of complications should also be taken into account. The appropriate diet should be prescribed along with symtomatic therapy (spasmolytics, analgesics, etc.)

Prophylaxis

Prophylactic measures are quite varied; this agrees with the variety of causes that provoke development of chronic colitis. Labour hygiene, sanitary conditions at home, and adequate nutrition are of primary importance. Patients with acute intestinal disorders should be thoroughly examined and treated. Regular out-patient observation of population is also important (control of intestinal parasitosis, treatment of constipation and other diseases of the digestive system that may cause pathologies in the large intestine, e.g. peptic ulcer, chronic gastritis, etc.)

MAIN CLINICAL SYNDROMES

Jaundice

Jaundice is an icteric colouration of the skin and mucosa by the increased content of bilirubin in the tissues and blood. The serum of blood taken from patients with true jaundice also becomes intense yellow. Jaundice is attended (often preceded) by changes in the colour of the urine, which becomes dark-yellow or brown; faeces can be very light or even colourless or on the contrary, dark-brown.

Jaundice can develop very quickly, within 1-2 days, to become very intensive, or it can develop gradually and be not pronounced (subicteric). Pantients themselves (or their relatives) notice yellow colour in their skin. They consult a adoctor for this reason. Jaundice can develop with severe itching of the skin, skin haemorrhages and haemorrhages of the nose and the gastro-intestinal tract.

Jaundice occurs in may disease of the liver, bile ducts, blood and also diseases of other organs and systems, to which bilirubin metabolic disorders are secondary. Some clinical symptoms attending jaundice are to a certain degree suggestive of its type and origin. Accurate diagnosis of various types of jaundice is possible with special laboratory studies.

True jaundice can develop due to the following three main causes: (1) excessive decompositon of erythrocyte and increased secretion of bilirubin (haemolytic jaundice); (2) impaired capture of unbound bilirubin by the liver cells and its inadequate combination with glucuronic acid (parenchymatous jaundice); (3) obstacles to exertion of bilirubin with bile into the intestine and reabsorption of bound bilirubin in the blood (obstructive jaundice).

Haemolytic (haematogenous) jaundice develops as a result of excessive destruction of erythrocytes in the cells of the reticulohistiocytic system (spleen, liver, bone marrow). The amount of unbound bilirubin formed from haemoglobin is so great that it exceeds the exertory liver capacity to account for its accumulation in the blood and development of jaundice. Haemolytic jaundice is the main symptom of haemolytic anaemia. It can also be a symptom of other diseases, such as B_{12}-(folic)-deficiency anaemia, malaria, protracted septic endocarditis, and other diseases.

The skin of a patient with haemolytic jaundice is lemon-yellow. Skin itching is absent. The amount of unbound bilirubin in the blood is moderately increased (50-200 per cent); the van den Bergh test for bilirubin is indirect. Bilirubin is absent from the urine but the urine is still coloured rather intensely by themarkedly increased (5-10 times) sterocobilinogen and (partly) urobilinogen. Faeces are intense dark due to the presence of considered amount of sterocobilinogen.

Parenchymatous (hepatocellular) jaundice develops due to the damage of the parenchyma cells (hepatocytes). These cells can capture bilirubin of the blood and bind it with glucuronic acid (The natural detoxicating function of the liver). The natural process of bilirubin excretion in the bile in the form of bilirubin glucuronide (bound (bilirubin) is thus impaired. The content of free and boudn bilirubin in the blood serum thus increases 4–10 times. In rare cases the increase may be even greater: free bilirubin increases due to hepatocyte dysfunction and bound bilirubin content increases as a result of back diffusion of bilirubin glucuronide from biliary into blood capillaries in dystrophy of the live cells. Bound bilirubin appears in the urine (bilirubin glucuronide is water soluble and easily passes via the capillary membranes as distinct from free bilirubin). Bile acids are also present in urine, but their content gradually increases. Excretion of stercobilinogen with faeces also decreases because the amount of bilirubin excreted by the liver into the intestine decreases, but faecs are rarely completely discoloured.

This type of jaundice is mainly determined by infection (virus hepatitis or Botkin's disease, leptospirosis) and toxic affections of the liver (poisoning with mushrooms, phosphorus, arsenic and other chemical substances, medicinal preparations included). But parenchymatous jaundice can develop also in liver cirrhosis.

The skin of patients with this jaundice is typically yellow with a reddish tint. Skin itching is less frequent than in obstructive jaundice because the synthesis of bile acids by the affected liver cells is upset. Symptoms of pronounced hepatic insufficiency may develop in severe course of the disease.

There exists a group of congenital pigmentary hepatoses in which the liver is not affected pathologically, the functional tests are negative, while the process of bilirubin conjugation with glucuronic acid is upset at some of these stages (Gilbert syndrome). This conditions is attended by a permanent or intermittent jaundice, which is sometimes pronounced and develops from infancy.

Obstructive(mechanical) jaundice develops due to parital or complete obstruction of the common bile duct. This occurs mostly due to compression of the duct from the outside, by a growing tumour (usually cancer of the head of the pancreas, cancer of the major duodenal papilla, etc.), or due to obstruction by a stone. Bile congestion above the point of obstruction develops and this elevates pressure inside bile passages in continuing bile excretion. As a result, the interlobúlar bile capillaries become distended and bile diffuses into the liver cells (where dystrophic processes develop) and passes into the lymph and the blood. Moreover, due to increased pressure inside fine bile capillaries, communications are formed at the periphery of the lobules between the capillaries and the lymph spaces, through which bile enters the blood vessels.

Skin and mucosa of patients with obstructive jaundice are yellow. Later, as bilirubin is oxidized to biliverdin, the skin and mucosa turn green and dark-olive. The bound bilirubin content in the blood with direct van der Bergh test is as high as 250–340 mmol/l or 15-20 mg/ 100 ml, and more. In protracted jaundice associated with liver dysfunction, free bilirubin content increases as well. Bound bilirubin can be found in the urine (the presence of bile pigments is determined by urinalysis) to give it brown colour and bright-yellow foaming. Faces are colourless either periodically (in incomplete obstruction, usually by a stone), or for lengthy periods of time (in compression of the bile duct by a tumour). Jaundice increases progressively in such cases; the skin and mucosa gradually turn greenish-brown; cachexia of the patient increases. In complete obstruction of the bile ducts, faeces become colourless (alcholic); their colour is clayish and grey-white; stercobilin is absent from faeces.

Bound bilirubin and also bile acids produced by the hepatocytes in ample quantity (cholaemia) are delivered to the blood in this type of jaundice. Some symptoms associated with toxicosis develop: pronounced skin itching, which intensifies by night, and bradycardia (Bile acids increase the tone of the vagus nerve by reflex). The nervous system is also affected: the patietn develops rapid fatigue, general weakness, adynamia, irritability, headache, and insomnia. If it is impossible to remove the cause of impatency of the common bile duct (stones or a tumour) the liver is gradually affected to add symptoms of hepatic insufficiency.

Portal Hypertension

Portal hypertension is characterized by a stable increase in the blood pressure in the portal vein. Protocaval anastomoses are dilated, ascites develops and the spleen increases in size.

Portal hypertension develops due to obstructed blood outflow from the portal vein as a result of its comperssion form the outside (by a tumour, enlarged lymph nodes of the porta hepatis in cancer metastases, etc.), or by obliteration of part of its intrahepatic branching in chronic affections of the liver parenchyma (in cirrhosis), or due to thrombosis of the portal vein of its branches. Growth and subsequent cicatrization of connective tissue at the site of degraded hepatic cells of a cirrhotic liver cause stenosis or complete obliteration of part of hepatic sinusoids and intrahepatic vessels. An obstacle is thus created to the blood flow which increases portal pressure and interferes with blood outflow from the abdominal viscera. In these conditions, transudatoin of fluids from the vessels into the abdominal cavity is intensified to account for the development of ascites. Decreased oncotic pressure of plasma is an important factor in the development of ascites associated with liver cirrhosis. The pressure decreases because of upset synthesis of albumins in the liver. Sodium and water retention is also important. It occurs due to hypersecretion of aldosterone by the adrenal glands (secondary *aldosteronism*) and its inadequate inactivation in the liver. The time of the onset of ascites depends on the degree of development of collateral circulation, i.e. on postocaval anastomoses. For a long time the disturbed portal circulation can be compensated for by delivery of blood into the superior and inferior venae cavae from the portal vein via normally existing anastomoses. But in portal hypertension these anastomoses become highly developed.

There exist three groups of natural protocaval anastomoses: (1) in the zone of haemorrhoidal venous plexus; these are anastomoses between the inferior mesenteric vein (the portal vein system) and haemorrhoidal veins emptying into the inferior vena cava; haemorrhoidal nodes develop in portal hypertension which rupture to cause rectal haemorrhage; (2) anastomoses in the zone of the oesophagogastric plexus: this is a collateral leading through the left gastric vein, the oesophageal plexus, and hemiazygos vein into the superior vena cava. In pronounced portal hypertension, marked varicose nodes are formed in the lower portion of oesophagus whose injury (e.g. by hard food) is responisble for possible haemorrhage in the form of haematemesis (blood vomiting), which is the most serious complication of diseases attended by portal hypertension and which is a frequent cause of death; (3) anastomoses in the system of paraumbilical veins communicating with the veins of the abdominal wall and the diaphragm, carrying blood to the superior and inferior venae cavae. In portal hypertension, varicose veins radiate from the umbilicus to give a peculiar pattern known as the caput medusae.

The degree of increase in the pressure in the portal vein system can be determined by a special needle and a water pressure gauge.

The pressure is measured in the spleen (splenometry) or in variocose veins of the oesophagus. In the latter case the needle is introduce through the oesophagoscope. It is believed that pressure in the spleen is the same as in the portal vein trunk. Normally it is 70-150 mm H_2O, while in portal hypertension it rises to 400–600 mm H_2O. Contrast techniques are used to reveal obstruction of the portal vein: these are splenoportography and in rare cases transumbilical portohepatography.

The spleen may be somewhat enlarged in venous congestion associated with portal hypertensin.

Treatment. In order to remove portal hypertension, whose first danger are *oesophagogastric* and *haemorrhoidal haemorrhages*, the patients are operated on for placing anastomoses between the portal vein system and the inferior vena cava.

Hepatolienal Syndrome

The *hepatolienal syndrome* is characterized by concurrent enlargement of the liver and the spleen in primay affection of either of these organs. Involvement of both organs in a pathological process (diseases of the liver, blood, certain infections, poisoning) is explained by their richly developed reticulohistiocytic tissue. In certain cases, e.g. in thrombosis of the hepatic veins, simultaneous enlargement of the liver and the spleen is determined by venous congestion in them. In addition to palpation, scanning can be used to reveal the hepatolienal syndrome.

Considerabale enlargement of the spleen is usually attended by its hyperfunction (*hypersplenism*), which is characterized by anaemia, leucopenia, and thrombocytopenia. The latter can cause haemorrhagic complications. These changes are explained by inhibitionof the haemopoiesis in the bone marrow due to hyperactivity of the spleen as a result of which destruction of the blood cells in the spleen is intensified, and antierythrocytic, antileucocytic, and antithrombocytic auto-antibodies are formed in the spleen.

Hepatic Insufficiency and Coma

Despite the considerable compensatory capacity of the liver, its grave acute and chronic diseases are attended by deep disorders in its numerous and very important functions due to the marked dystrophy an destruction of the hepatocytes. Clinicists define this condition as the hepatic insufficiency syndrome.

Depending on the character and acuity of the affection acute and chronic hepatic insufficiency are distinguished. The following three stages of the disease are also distinguished: (1) early compensated stage; (2) pronounced decompensated; and (3) terminal dystrophic stage that ends in a hepatic coma and death.

Acute hepatic insufficiency arises in grave forms of virus hepatitis (Botkin's disease) and poisoning with hepatotropic substances (affecting the live in the first instance). These may be chemical substances (e.g. phosphorus compounds, arsenic, large doses of alcohol) or vegetable poisons (inedible mushrooms containing amanitotixin, helvellic acid, or muscarien extracted of male fern, etc.). Acute hepatic insufficiency develops rapidly, within seveal days or hours.

Chronic hepatic insufficiency develops in many chronic diseases of the liver, e.g in cirrhosis and tumours. Its development is slow and gradual.

Development of hepatic insufficiency is underlain by marked dystrophy and necrobiosis of hepatocytes which is attended by a considerable impairment of all liver functions with formationof collaterals between the portal vein system and the venae cavae. Collaterals develop in cases when the blood flow from the portal vein into the liver is obstructed in any affection of this organ. Large amounts of blood containing toxic substancs absorbable in the large intestine pass through the collaterals into the greater circulation system to bypass the liver. Hepaptic insufficiency is explained by various complicated metabolic disorders in the liver, upset bile secretory and excretory function, and impaired detoxicating function of the liver.

The pathogenesis of hepatic coma is manifested by grave self-poisoning of the body due to almost complete dysfunction of the liver. The body is poisoned by the non-detoxicated products of intestinal (bacterial) protein decomposition, final products of metabolism, and especially ammonia. Normally, the major part of ammonia is captured by the hepatocytes and converted to urea (in the ornithine Kreb's cycle), which is then excreted by the kidneys. Phenols, which are normaly detoxicated in the liver by their combination with glucuronic and sulphuric acids, also have a toxic effect. Other toxic substancaes also accumulate in the blood in the presence of hepatic insufficiency. The electrolyte metabolism becomes upset, and in severe cases, hypokaliaemia and alkalosis develop.

Hepatic insufficiency may be aggravated and coma may be aggravated and coma may be provoked by alcohol, barbiturates, some analgesics (morphine, promedol), protein0rich diet (which intensifies putrefactive processes in the intestine, production of toxic substances and their absorption in the blood), by profuse haemorrhage from the digestive tract (which often aggravates portal cirhosis of the liver), by large doses of diuretics, instantansous withdrawal of large amoutns of ascitic fluid, severe diarrhoea, and the attending grave infectious diseases.

Clinical signs of hepatic insufficiency usually combine with symptoms of the liver disease that provoked hepatic insufficiency.

The intensification of the symptoms by stages is vividly illustrated by the progressive development of hepatic insufficiency (in patients with liver cirrhosis, tumour of the liver, and other diseases of this organ).

The clinical symptoms are absent during the early stage of hepatic insufficiency, but the body's tolerance to alcohol and also to other toxic substances decreases, and the findings of laboratry load tests change.

During the second stage, clinical signs of hepatic insufficiency develop; first mild but later more pronoucned non-motivated fatigue, poor appetite, increased weakness in usual physical exertion, frequent dyspepsis (poor tolerance of fat food, the presence of meteorism, rumbling and pain in the abdomen, changed stools), which are explained by disorders of bile secretion and digestive processes in the intestine. Upset assimilation of vitamins explains polyhypovitaminosis. Fever, which is not infrequent in hepatic insufficiency, can be due to both the main disease and impaired detoxication of some proteinous pyrogens by the liver. Jaunndice and hyperbilirubinaemia with accumulation of free (indirect) bilirubin in the blod are frequent in hepatic insufficiency. Deranged structure of the liver and upset cholestasis can stimulate accumulationof bilirubin glucuronide ("direct" bilirubin) in the patient's blood.

Deranged albumin synthesis in the liver and also pronounced hypoalbuminaemia can cause hypoproteinaemic oedema and intensify ascites, which often occurs in patients with chronic liver affections. Upset synthesis of some blood coagulating factors (ribrinogen, prothromin, proconvertin) and also decreased blood platelet content (due to hypersplenism that attends many chronic diseases of the liver) provoke the onset of haemorrhagic diathesis (skin haemorrhages, nasal bleeding, haemorrhage in the intestinal tract). Inadequate inactivation of oestrogens by the chronically affected liver provokes endocrine disorders (gynaecomastia in men, menstrual disorders in women, etc.).

Changes in laboratory tests (hepatic tests) are significant in the second stage of hepatic insufficiency. Characteristic is the decreased content of substances produced by the liver: albumin, cholesterol, fibrinogen, et. Considerable changes in the liver function are also revealed by radioisotope hepatography.

The third, final stage of hepatic insufficiency is characterized by even deeper metabolic disorders and dystrophic changes, which are pronounced not only in the liver but also in other organs. Patients with chronic liver disease develop cachexia. They also suffer from nervous and psychic disorders whichare precursors of coma; decreased mental ability, slow thinking, slight euphoria, sometimes depression, and apathy. The patient becomes easily irritable, his moods are quickly changed, attacks of melancholy and frustration occur at times, and sleep is deranged. Derangement of consiousness and loss of orientation in time and space develop along with partial loss of memory, disorderded speech, hallucinations, and somnolence. Specific tremor (slow and fast) of the upper and lower limbs is characteristic.

The precoma pepriod may last from a few hours to seveal days and even weeks. The patient may recover from this state, but coma develops in most cases.

The clinical picture is characterized first by excitation and then by general inhibition (stupor) and progressive derangement of consciousness (sopor), to its complete loss (coma). The ECG curve is flattened. The reflexes are decreased, but hyper-reflexia and pathological reflexes (sucking and grasping) develop. Motor anxiety, clonic convulsions due to hypokaliaemia, muscular twitching , and tremor of the extremities (arrhythmical and rhythmical twitching of the fingers and toes) are characteristic. Respiratoin rhythm becomes upset. Kussmaul respiration (less frequently Cheyne-Stokes respiration) develops. Incontinence of faeces and urine ensues. The patient's breath (and also urine and sweat) smells "sweety hepatic" (fetor hepaticus) because of liberation of methyl mercaptan which is formed in deranged methionine metabolism. Inspection of the patient often reveals signs of haemorrhagic diathesis (bleeding gums, nasal and skin haemorrhage). The patient's temperature in the terminal period is subnormal. Jaundice is intensified. The liver may remain enlarged or its size may decrease. Laboratory tests show moderate anaemia, leucocytosis, increased ESR, low counts of platelets and fibrinogen; the prothrombin time increases, hepatic functional tests become sharply upset, and the bilirubin level increases. The content of rsidual nitrogne and ammonia in the blood serum increases to indicate secondary affections of the kidneys (the hepatorenal syndrome). Hyponatriaemia, hypokaliaemia, and metabolic acidosis develops. Heaptic coma usually terminates fatally. But the patient can in some cases be saved.

Treatment

Intensive therapy is required in acute hepatic insufficiency: infusionof plasma, polyglucin, glutamic acid solution (to bind ammonia), oxygen, correction of waster water-salt disorders; the life of the patient should be maintained during the critical period (several days) to help the strong regenerative capacity of the live. In addition to the treatment of the main disease, chronic liver diseases should be tgreated by removing the aggravating factors (oesophagogastric haemorrhage, attending infections). The diet should be poor in protein in order to suppress the putrefactive processes in the intestine; antiobiotics decrease absorption of decomposed proteins. The electrolyte disorders are corrected and haemorrhages are controlled.

Recently methods for transplantation of healthy live to patients with hepatic insufficiency are being developed.

SEPECIAL PATHOLOGY

The most common affection of the live is its inflammation. Acute and chronic hepatitis, and also cirrhosis and hepatosis are distinguished. Primary cancer of the liver occurs in rare caes; as a rule malignant tumours metastasize into the liver from other organs. Echinococcus usually localizes in the live. The live is also affected in opisthorchiasis and some other parasitary invasions.

Affections of the bile secretion system are quite common. These are cholelithiasis, acute and chronic cholecystitis, cholangitis, dyskinesia of the bile ducts, and primary cancer of the gall bladder.

Chronic Hepatitis

Chronic hepatitis is a chronic diffuse or focal inflammatory affection of the live.

Aetiology

The following groups of chronic hepatitides are distinguished: (1) infectious and parasitogenic; infectious hepatitis develops secondary to virus hepatitis, brucellosis, tuberculosis, syphilis, and some other diseases;; (2) toxic hepatitis caused by industrial, medicamentous, domestic and food chronic poisoning by hepatotropic toxic substances (chloroform, trinitrotoluene, aminazine, lead compounds, etc.); (3) toxico-allergic hepatitis, which develops not only in response to direct toxic effect of some medicines or hepatotropic chemicals, but also due to hypersensitivity of the liver cells of the entire body to these substances (medicamentous hepatitis, hepatitis associated with collagenosis); (4) metabolic hepatitis, which arises due to metabolic disorders in the live, associated withprotein-vitamin deficiency, and also in fat dystrophy and amyloidosis.

In 40-70 per cent of cases chronic hepatitis develops as an outcome of an acute epidemic or serum hepatitis. Hepatitids are mostly diffuse affections. Liver affections ar focal in tuberculosis (tuberculous granuloma, caeous abscesses or tuberculoma), symphilis (gumma), some protozoal disease (amoebic abscesses), fungal and bacterial affections (usually abscesses), and in some other cases.

Pathogenesis

This is mostly determined by the aetiology of the disease. When exposed progressively affected (To necrobiosis); the secondary inflammatory reaction of the liver mesenchyma is equally important in the pathogenesis of chronic hepatitis. Hepatitis of virus nature is probably associated with peristence of the virus in the liver cells and with the progressive cytopathic effect of this virus, which kills the hepatocytes to cause inflammatory reaction of the connective tissue. In many cases autoimmune processes are of primary importance. They arise in response to the primary affection of the liver tissue by any aetiological factor. Obstructed bile excretion and bile congestion, cholangitis and cholangiolitis (with subsequent extension of inflammation onto the liver tissue), and also some medicamentous poisonings (phenothiazine derivative) are decisive in the pathogenesis of the so-called cholestatic hepatitis.

Pathological anatomy

Among diffuse inflammatory affections of the liver benign (non-active, persisting), aggressive (recurrent, active), and cholsestati chronic hepatitis are distinguished.

Non-active hepatitis is characterized by inflammation in the periportal zones, preservation of the lobular structure, and sometiems by moderate dystrophic changes in the hepatocytes. The liver is enlarged, the capsule is thickened, and streaks of connective tissue are seen on its surface. The inflammatory and cicatricial processes are more distinct in the liver affected by active hepatitis. Inflammatory infiltration extends from the periportal zones inside the liver lobules whose outlines are indistinct. Hepatocytes are extensively necrotized and have dystrophic changes; fibrosis is found in the liver. The size of the liver increases, its surfaces are coarse and necrotized zones can be seen as red amorphous spots. Cholestatic hepatitis is also characterized by marked affection of bile ducts (cholangitis and cholangiolitis) and signs of cholestasis.

Hepatocellular hepatitis (epithelial and parenchymatous) and mainly mesenchymal hepatitis are also differentiated.

Clinical picture

Chronic hepatitides are characterized by (1) dyspeptic symptoms; (2) jaundice (it may be absent in some cases); (3) moderate enlargement and induration of the liver and the spleen; (4) dysfunction of the liver as determined by laboratory tests and radiohepatography. But the clinical picture and also the course of each clinico-morphological form of hepatitis have their special features.

Chronic benign hepatitis is characterized by obliterated clinical picture. The patients complain of heaviness or dull pain in the right hypochondrium, decreased appetite, bitter taste in the mouth, nausea and eructation. Jaundice is usually absent or its is moderate. Objective studies reveal a mildly enlarged liver with a smooth surface and a moderately firm edge, which is slightly tender to palpation. Enlargement of the spleen is not marked.

Laboratory studies

The blood bilirubin content is usually normal; in the presence of jaundice it increases to about 17-50 μmol/l (1–3 mlg/100 ml); the blood globulin content is mildly increased, avtivity of the enzymes is either normal or only slighly changed; th prothrombin content is normal or slightly decreased; the bromsulphthalein test is slightly positive.

Chronic active (aggressive) hepatitis is characterized by complaints and objective symptoms; weakness, loss of weight, fever, pain in the right hypochondrium, loss of appetite, nausea, regurgitation, meteorism, skin itching, jaundice, and frequent nasal bleeding. The liver is enlarged, firm, with a sharp edge. The spleen is enlarged.

Laboratory tests often reveal anaemia, leucopenia, thrombocytopenia (a sign of hypersplenism), and increased ESR. Functional tests are changed cosisereably: they show hyperbilirubinaemia, hyperproteinaemia, hypergammaglobulinaemia, positive protein-sedimentation tests, increased activity of transaminase, aldolase, and alkaline phosphotases; decreased activityof cholinesterase. The serum iron content is significantly increaed while the prothrombin index is sharply decreased; excretion of bromsulphthalein is delayed.

Puncture biopsy of the liver and (for special indiations) laparoscopy establish the special histologicl and macroscopic changes in the live characteristic of these forms. These techniques are also used for differentiation of chronic hepatitis from other diseases of the liver (cirrhosis, amyloidosis, etc). It should be noted that histological and histochemical studies of liver bioptates often reveal early morphologicla changes in the liver which precede the clinical and laboratory signs of chronic hepatitis.

Chronic cholestatic hepatitis is mainly characterized by the cholestatic syndrome; jaundice (subhepatic), severe skin itching, hyperbilirubinaemia, increased activity of alkaline phosphatase in the blood, and high cholesterol of blood. Persistent subfebrile temperature and regular increase in ESR are also not infrequent.

Course

A benign persistent chronic hepatitis can last to 20 years; exacerbations are rare and arise only in the presence of strong provoking factors. Liver cirrhosis develops in rare cases. Complete clinical recovery is sometimes possible, especially so if the patient is specifically treated. The morphological structure of the liver is restored in such cases.

Aggressive hepatitis is characterized by relapses, whose frequency depends on various factors. Frequent relapses accelerate progressive dystrophic, inflammatory and cicatricial changes in the liver and stimulate development of cirrhosis. Prognosis in this form of hepatitis is bad.

The course and prognosis is cholestatic hepatitis depend on its aetiology and the possibility of removing the obstacle of bile outflow (in compression of the common bile duct by a tumour, cicatrical or inflammatory stenosis, etc.).

Treatment

The cause of chronic hepatitis should be removed in the first instance: complete discontinuation of taking alcohol or exposure to harmful substances, etc.

During exacerbations, the liver should be spared as much as possible and regenertion of liver cells stimulated (Bed rest, diet, parenteral use of vitamins, glucose, etc.)

Cirrhosis of the Liver

Cirrhosis of the liver is a chronic progressive disease characterized by increasing hepatic insufficiency in connection with dystrophy of the liver cells, cicatricial cirrhosis, and structural reconstruction of the liver.

Aetiology

Cirrhosis of the liver is a polyaetiological disease. It may develop due to (1) infection (virus of epidemic hepatitis); (2) alcoholism; (3) protein- and vitamin-deficient diet; (4) toxico-allergic factor; (5) cholestasis. Of the mentioned aetiological factors the leading role in this country belongs to the virus of epidemic hepatitis. Cirrhosis caused by the virus is probably explained by its long persistence in the liver cells.

Chronic alcoholic poisoning is also a very important aetiological factor. It affects absorption of vitamins and proteins in the intestine to provoke cirrhosis of the liver. It also acts directly and specificially on metabolism of the liver cells. The alimentary factor (malnutrition, mainly protein-and vitamin deficit) is a frequent cause of liver cirrhosis in some developing countries. In this country the alimentary factor (malnutrition) is only of endogenous origin: deranged absorption of proteins and vitamins (in grave chronic diseases of the gatro-intestinala tract, in patients with totala resection of the stomach, resection of the intensive, chronic pancreatitis, and in some other cases). Toxic cirrhosis of the liver arises in repeated and chronic exposure to carbon tetrachloride, compounds of phosphorus or arsenic, in food poisoning (inedible mushrooms, seeds of heliotrope). Toxico-allergic cirrhosis of the liver includes also affections connected with hypersensitivity (autoallergy) to various drugs (Aminazine, chloroform, some antibiotics, sulpha preparations, etc); hypersensitivity can cause dystrophy and necrosis of the liver parenchyma.

Obturation of intra- and extrahepatic bile ducts and their inflammation cause congestionof bile and cholestasis,k and are important factors in the development of biliary cirrhosis.

The aetiological factors does not always determine the way of development of liver cirrhosis. One and the same factor can cause various morphological variants of cirrhosis (portal, postnecrotic, and biliary); at the same time various aetiological factors can cause similar morphological changes.

Pathogenesis

The pathogenesis of liver cirrhosis is closely connected with morhogenesis. The greatest importance in the developmental mechanism of liver cirrhosis belongs to recurrent necrosis of the liver cells which is provoked by aetiological factors and cause collapse of the reticulin fraamework of the liver, formation of cicatrices, and derangement of circulation in the adjacent portions of the preserved liver parenchyma.

Intact hepatocytes or lobe fragments begin their intense regnearation under the effect of growth stimulants supplied form the necrotic focus. The formed large nodes of regenerated tissue compress the surrounding tissue with the invested vessels; thehepatic veins are compressed especially strongly. The blood outflow becomes upset to provoke portal hypertension and formation of anastomoses between the branches of the portal and hepatic veins that facilitate intrahepatic circulation. Blood now bypass the liver parenchyma to impair drastically its blood supply, to cause new ischaemic necroses, and to stimulate the progress of cirrhosis even in the absence of the primary aetiologicla factor. Collagenous connective tissue grows intensively: connective tissue partitions (septa) grow into the parenchyma from the periportal fields to cause fragmentation of the liver lobules. These false lobules can later become the source of nodular regeneration. Chronic direct exposure to certain toxic hepatotropic substances, and also autoimmune and some other mechanisms are important in the pathogenesis of certain forms of cirrhosis.

Pathological anatomy

Three main morphological variants of liver cirrhosis are distinguished: portal (septal), postnecrotic, and biliary.

Portal (septal) cirrhosis of the liver is usually the result of alimentary insufficiency and alcoholism; less frequently it is secondary to Botkin's disease (viruss hepaptitis). Its development is underlain by formation of connective-tissue septa interconnecting periportal fields, with the central zone of the lobule and causing its fragmentation. Macroscopically the liver may be enlarged or diminished. Small nodes of regenerated tissue, circumscribed by narrow septa of connective tissue, are distributed uniformly over the entire surface of the live. The nodes are almost equal in size. The microscopic picture: marked fatty infiltration of the liver cells is observed in alimentary or alcoholic cirrhosis; these changes may be absent in cirrhosis that develops after virus hepatitis. "False bile ductules", leucocyte infiltration, and compressed small veins are found in the stroma between the nodes of regenerated tissue.

Postnecrotic cirrhosis of the liver develops as a result of submassive and masive necrosis of theliver cells due to virus and (less frequently) toxic hepatitis. Macroscopic picture is characterized by irregular changes in the liver, which is usually diminished in size. Nodes of various form and size can be seen on the liver surface. Microscopy shows irregular nodes of regenerated tissue and intact portions of the parenchyma. Broad fields of collapsed collagenized stroma with closely running portal tracts, venules, and cell infiltrates can be found between the nodes of the regenerated tissue. Inflammatory infiltration is marked.

Biliary cirrhosis of the liver has two variants. Primary biliary cirrhosis (pericholangiolitic) arises after epidemic hepatitis or toxico-allergic action of some medicinal preparations. Its development is underlain by obstruction of fine intrahepatic bile ductules which accounts for bile congestion. Macroscopy: the liver is enlarged and consolidated and is dark-green or olive in colour; it is microgranular. Extrahepatic bile ducts are patient. Microscopy is characterized by the presence of intralobular and perioportal cholestases. The periportal fields are broad, with fibrosis around proliferating cholangioles; intralobular fibrosis develops around intralobular cholangioles with dissociation of the liver cells and their groups. Secondary biliary cirrhosis arises as a result of prolonged obstruction of extrahepatic bile ducts by stones, tumour, etc. It provokes dilation of the bile ducts, development of cholangitis, and pericholangitis; cirrhosis of the liver develops if these changes progress.

In addition to the described variants of cirrhosis, they may also be mixed: morphological signs of other variants may join the main variant. Activity of cirrhosis is characterized by the presence of new dystrophic and regenerative processes in the parenchyma, intense inflammatory infiltration in the stroma, proliferation of cholangioles, indistinct borders between nodular parenchyma and internodular stroma. A neglected cirrhotic process is characterized by replacement of liver tissue by nodes of regenerated tisisue, markedly pronounced portal hypertension, large quantity of vascular connective-tissue septa growing into the parenchyma (portohepatic anastomoses). According to the morphological picture, fine-and large-nodular cirrhosis is distinguished. Mixed variants also occur.

Clinical picture

Portal cirrhosis of the liver occurs mostly between the ages of 40 and 60. The incidence in men is twice higher than in women. Postnecrotic and biliary cirrhosis of the liver develop in younger patients, mostly in women.

Clinical manifestations of liver cirrhosis depend on the degree of affection of the liver cells and the associated hepatic dysfunction and portal hypertension, on the stage of the disease (compensated or decompensated), and also on the activity of the process. The following symptoms of the disease are most characteristic of the majority of patients with various forms of liver cirrhosis.

Pain in the region of the liver, in the epigastrium, or diffuse pain in the whole abdomen is usually dull and boring, intensifying after meals, especially after fatty food, ample drinking and physical exercise. Pain is usually associated with enlargement of the liver and distension of the capsule, or with necrotic foci located near the capsule, with perihepatic symptoms, and also concurrent inflammatory affections of the bile ducts.

Dyspepsia in the form of decreased appetite to complete anorexia, the feeling of heaviness in the epigastrium after meals, nausea, vomiting, meteorism and dyspeptic stools (especially after fatty meals) depend mainly on deranged secretion of bile and hence defective digestion. But they can also be associated with the attending dyskinesia of the bile ducts or alcoholic gastroenteritis.

Decreased work capacity, general weakness, fatigue and insomnia are often observed in cirrhosis of the liver. Fever is usually irregular and sometimes of the undulant type. It often attends postnecrotic cirrhosis of the liver and is explained by necrotic destruction of the liver cells. Marked fever is characteristic of the active period and infectious cirrhosis.

A haemorrhagic syndrome is observed in 50 per cent of patients with cirrhosis of the liver. Profuse bleeding from varicose veins of the oesophagus and the stomach can often be early signs of portal cirrhosis; they are caused by increased pressure in the veins of the oesophagus and the stomach. In other variants of cirrhosis nasal, gum, uterine and skin haemorrhages develop in marked decompensation. They depend on the decreased coagulability of blood due to liver dysfunction.

Signs of cirrhosis are as follows. Cachexia is especially characteristic of patients with portal cirrhosis of the liver. In long-standing disease the subcutaneous fat disappears along with atrophy of muscles, especially of the upper shoulder girdle. The appearance of such patients is quite specific: the face is very thin with grey or subicteric skin; the lips and the tongue are bright-red; the cheek bone region is affected by erythema; the extremities are thin and the abdomen is large (due to ascites, enlarged liver and spleen); the subcutaneous veins of the abdominal wall are dilated, the legs are oedematous. Malnutriton is usually associated with disordered digestion and assimilation of food, and impaired synthesis of proteins in the affected liver.

Jaundice (Except the caes with biliary cirrhosis) is a sign of hepatocellular insufficiency associated with necrosis of the liver cells. The affected hepatocytes partly lose their capacity to capture bilirubin from blood and to bind it with glucuronic acid. Bilirubin excretion into bile is disordered as well. Free (indirect) and bound (direct) bilirubin of blood serum therefore also increases. Jaundice is usually characterized by partial decolouration of faeces and by the pressence of bile in the duodenal contents. Jaundice is often attended by skin itching. Jaundice associated with biliary cirrhosis resembles obstructive jaundice; severe skin itching is observed. The intensity of jaundice varies from light subicteric to marked jaundice (depending on the degree of obstruction of the bile ducts).In prolonged obstruction of the extrahepatic duct the skin acquires a greenish tint which depends on oxidation of bilirubin to bilverdin. Moreover, brown pigmentation of the skin may also be observed. It depends on accumulation of melanin.

"Minor" signs of cirrhosis can be revealed during examination of the patient. These signs are as follows: (1) spider angiomata (they may develop years before marked symptoms of the disease develop); their number increases and the colour intensifies during exacerbation of the disease; (2) erythema of the palms; (3) red lustrous lips, scarlet mucosa of the mouth, scarlet (lacquered) tongue; (4) gynaecomastia (increased mamary glands) and other female sex characters developing in men (decreasing growth of the hair on the face, chest, abdomen, and the head); (5) xanthomatous plaques on the skin (observed in patients withi biliary cirrhosis of the liver); (6) Hippocrates fingers with hyperaemic skin at the nail beds. Inspection of the abominal skin can reveal dilation of the veins that can be seen through the thinned skin of the abdominal wall (caput medusae). Collateral venous system can be seen on the chest as well. Haemorrhoidal veins are often dilated.

Ascites is the most characteristic sign of portal cirrhosis. Ascites may develop slowly and the abdomen grow to huge size; the patient develops dyspnoea. Oedema may develop; hydrothorax may also occur in some cases. In other variatns of cirrhosis, ascites develops at later stages of the disease.

Enlarged liver can be palpated in 50-75 per cent of patitns with cirrhosis. The enlargement can be insignificant, only determinable by percussion,or considerable when the liver occupies the entire left part of the abdominal cavity. The liver is firm, the surface is sometimes irregular, and the lower edge sharp. Enlargaement of the spleen is often attended by its increased activity (hypersplenism).

Laboratory findings

An active cirrhotic process is characterized by anaemia,

leucopenia, thrombocytopenia, and increased ESR. Anaemia can be due to hypersplenism and gastro-intestinal haemorrhage, hepatocellular insufficiency, and often increased heamolysis, whichis accompanied by reticulocytosis of the peripheral blood.

The blood serum bilirubin content becomes considerable only in the final stage of the disease. At the same time, the affection of the excretory function of the cirrhotic liver can be assessed by the presence of the conjugated fraction of bilirubin (bound bilirubin). Its content increases in normal an dincreased total bilirubin. The free bilirubin content increases in the blood serum as a result of upset conjugation of bilirubin in the liver cell and haemolysis. The blood serum bilirubin content varies in biliary cirrhosis of the liver from 26 to 340 μmol/l (1.5–20 mg/100 ml), mostly at the expense of boudn bilirubin.

The presence of much urobilin in the urine indicates liver insufficiency. The amount of urobilin in the urine and stercobilinin the faces decreases in the presence of pronounced jaundice when a small amount of bilirubin enters the intestine. Bilirubin is found in the urine of patients with jaundice.

The upset excretory function of the liver is manifested by retnetion of bromsulphthalein in the blood (during its intravenous administration) and also by radioisotopic hepatography and scanning of the liver.

Affection of liver cells is manifested by characteristic changes in the protein indices: decreased concentration of serum albumins and hypergammaglobulinaemia which in turn decreases the albumin-globulin coefficient. Activation of the inflammatory process in the liver involves an increase in the a2-globulins, while jaundice causes an increase in b-globulins. During remissions, all these changes become less pronounced. The blood level of lipids and cholesterol also increases considerably in the presence of biliary cirrhosis. A sensitive index of liver dysfunction is the decreased activity of cholinesterase. Transaminase activity increases in exacerbation of liver cirrhosis. Activity of alkaline phosphatase also increases in biliary cirrhosis.

The decreased prothrombin content (which is synthesized by the liver cells), increased antithrombin coagulative activity and decreased total coagulative activity of plasma are imprtant in the aetiology of haemorrhagic diathesis in liver cirrhosis.

Laparoscopy and especially biopsy of the liver help reveal intrravital morhological signs of each variant of liver cirrhosis. Varicose veins of the oesophagus are revealed by X-rays.

It is not always possible to differentiate between all variants of liver cirrhosis from the data of clinical and instrumetnal methods of examination, nevertheless, by comparing the mentioned signs, one can notice that the symptoms of portal hypertension in portal cirrhosis of the liver are often revealed long before the functional insufficiency develops. Hepatic insufficientcy only develops at a later stae of the disease. But in the presence of postnecrotic cirrhosis of the liver, the symptoms of hepatic insufficiency develop early. They largely determine the entire clinical picture of the disease. Chronic jaundice (obstructive type) prevails in the clinical picture of biliary cirrhosis along with satisfactory general condition of the patient, who suffers from skin itching, sometimes fever (associated with chills); the blood alkaline phosphatase and cholesterol content increases. Transcutaneous cholangiography is used to determine the cause of cholestasis. The procedure is done when indicated.

Complaints of patients with *compensated liver cirrhosis* are not serious. The disease is often revealed accidentally durign examination (enlarged liver and spleen). Remissions may be long (measured by years). Decompensated active cirrhosis is characterized by market symptoms of the disease and rapid progressive course.

Course

The course of the disease is usually progressive. The overall term of the disease is usually 3 to 5 years; in rare cases the disease may last 10 years and even longer (usually in biliary cirrhosis of the liver).

The terminal period of the disease, irrespective of the form of cirrhosis, is characterized by gastro-intestinal haemorrhage and progressive signs of functional insufficiency of the liver, with finally developing coma. These are two most frequent direct causes of death of patients with liver cirrhosis. Gastro-intestinal haemorrhage (blood vomiting and melaena) is caused by the rupture of varicose nodes in the lower third of the oesophagus or, less frequently, in the stomach. A direct cause of varicose haemorrhage is physical strain or local affection of the mucosa (e.g. by coarse food). Profuse haemorrhage (if it does not cause death) can cause anaemia with subsequent impairment of the function of the liver cells and accelerated development of heaptaic coma.

Treatment

Cirrhosis of the liver in the compensation stage is treated by preventing its further affection with alcohol, toxic substances, etc., and also by rational organization of work regimen and nutrition (high-calories diet rich in protein and vitamins). During decompensation stage, hospital treatment is required. Glucocorticosteroid hormones are given in the active process (except cases complicated by dilation of the oesophageal veins); syrepar (hydrolysate of cattel liver), essential (a complex preparation containing essential phospholipids), and vitamins are also prescribed. Patients with ascites are prescribed a diet restricted in salt and diuretics (periodicaly). If ascites cannot be cured by diuretics, the fluid is released by paracentesis.

In order to decrease the lipid content of the serum in primary biliary cirrhosis, lipoic acid preparations are presribed. Skin itching is removed by chollestyramin e(preparation binding fatty acids). Surgical treatment is indicated in cases with secondary biilary cirrhosis of the liver, e.g. in obstruction of the bile duct by a stone.

Cholelithiasis

Choleliithiasis is characterized by formation of stones in the gall bladder or, less frequent, in the bile ducts. The incidence of the disease is rather high. According to the data of postmortem examination, stones are found in the gall bladder of every tenth patient who dies from various causes. At the same time, clinical signs of cholelithiasis are only found in 10 per cent of carriers of stones, mainly in women aged 30-55.

Aetiology and pathogenesis

The disease is underlain by general metabolic disordrs which provoke formation of stones. Infection and bile congestion are also important. Upset cholesterol metabolism with hypercholesterolaemia attended by the increased bile cholesterol content is decesive because most stones contain cholesterol. This is also confirmed by the fact that cholelithiasis often concurs with atherosclerosis, diabetes mellitus, obesity, and othe rconditions attended by hypercholesterolaemia. Frequent formation of pigment stones contain cholesterol. This is also confirmed by the fact that cholelithiasis often concurs with atherosclerosis, diabetes mellitus, obesity, and other conditions attended by hypercholesterolaemia. Frequent formationof pigmented stones in the presence of excess billirubin in bile in haemolytic anaemia (haemolytic jaundice) is explained in a similar way. At the same time, blood cholesterol is not increased in all patients with cholelithiasis. There is no parallelism between cholesterol level of bile and blood.

The main components of stones are cholesterol, bilirubin, and calcium. These are contained in bile in the form of unstable colloidal solutions. Cholesterol is retained in bile mainly be bile acids. When cholate level in the bile decreaes, cholesterol precipitates as crystals. The ratio between salts of bile acids (cholates) And cholesterol in normal bile is 15:1; in cholelithiasis this ratio decreaes to 6:1. Disorders in the physio-chemical composition of bile are believed to be decisive for stone foromation. Hepatocyte dysfunction decreases the formation of bile acids which may be the cause of dyscholia.

The importance of the infectious factor consists in that protein-rich exudate of inflamed gall bladder upsets the normal colloidal and chemical composition of bile to precipitate bilirubin, cholesterol and calcium, and to cause formation of mixed stones typical for infectious diseases of the gall bladder.

Bile congestion in the gall bladder provides conditions for stone formation because it promotes concentrationof bile and stimualtes an increase (10-12 times) in cholesterol concentration in the bile, while gradual absorption of bile acids decreases their content in the bile. Moreover, bile congestion can provide favourable conditions for development of infection. Disordered neurohumoral regulation of contractility of the gall bladder and bile ducts (dyskinesia), as well as the anatomical changes in the bile passages (bends, adhesions, scars) are essential factors provoking bile congestion. Factors interfering with normal emptying of the gall bladder are alsi important: increased intra-abominal pressure (e.g. during pregnancy), ptosis of the internal organs, persistent constipations, hypodynamia, and rare meals are among these factors.

Hereditary predisposition is also very important. Stones often occur in several generations of one family (especially among women). Excessive food rich in fats and calories causes hypercholesterolaemia and stimulates formation of gall stones.

Pathological anatomy

There exist three major groups of gall stones. Purely cholesterol stones are white or yellowish concretions which are found in the gall bladder. They usually develop as single round or oval concretions. The stones are light (float on the surface of water) and burn with a bright flame. The section of a stone shows its radial structure (crystals of cholesterol). Pigment stones consist of bilirubin and calcium. Their shapes vary. Usually pigment stones are small and multiple. Their colour is black with a greenish tint; the stones are dense but brittle. Stones consisting of only calcium carbonate rarely occur. Mixed stones consisting of cholesterol, calcium, and pigment occur most frequently. They are heavier than water and burn with difficulty. A section reveals laminar structure. The shape and size of mixed stones vary but they are usually small and multiple. If stones are tightly packed in the gall bladder, their surfaces become facetted (from mutual pressure).

The mucosa of the gall bladder can be affected with inflammation in the presence of stones. Prolonged presence of stones in a non-inflamed gall bladder can cause atrophy and sclerosis of the bladder wall or (in very rare cases) decubitus and perforation of its walls.

Clinical picture

Pain attacks in the right hypochondrium (the so-called biliary colic) are the most characteristic symptom of cholelithiasis. Colics are usually provoked by small stones as they move in the region of the neck of the gall bladder, its isthmus, or directly in the cystic duct. Pain is caused by spastic contractions of the gall bladder and the ducts which develop as result of a sudden distension of the gall bladder and increased pressure inside it due to a mechanical obstruction to bile outflow. The pain develops also by the reflex mechanism, as a response to irritation of the cystic duct receptors by stones. A gall bladder colic can be provoked by physical or nervous strain, jolting motion, ingestion of much fat, etc.

A gall-stones colic develops suddenly. The pain is first diffuse and is felt in the entire right hypochondrium. Later it localizes in the region of the gall bladder or in the epigastrium. Pain is piercing and so severe that cannot be tolearated without pain-relieving preparations. The patient groans and tosss in bed vainly seeking for a convenient posture. The pain can specifically radiate upwards, to the right posteriorly, to the right shoulder, neck, the jaw and into the right subscapular region. Pain can radiate also into the heart to provoke an attack of angina pectoris.

Pain can continue from several minutes to a few hours and even days, periodically subsiding and strengthening. Intensified contractions of the gall bladder promote further propulsion of the stone (not more than 1.5 cm in diameter) from the neck or the cystic duct into the common bile duct. Sometimes, after relaxation of the spasm the stone may return back to the "silent" zone (the fundus of the gall bladder). In both cases, the pain attack discontinues as suddenly as it begins. The patient's condition rapidly improves. The attack can often be alleviated by applying warmth or giving spasmolytics (atropine, sulphate, 16 ml of a 0.1 per cent solution, or papaverine hydrochloride, 2 ml of a 2 per cent solution subcutaneously). This is a valuable differential-diagnostic sign: these remedises fail to relieve pain in acute cholecystitis, while warmth (e.g. a hot-water bottle) is contraindicated because warmth intensifies blood inflow and the inflammatory processes.

If the colic is long standing, jaundice may develop at the end due to a spasm of the common bile duct. The jaundice usually is not intense and is only transient (2 to 3 days).

Gall-stone colic is usually attended by nausea and recurrent vomiting. The reflex mechanism explains the fever which often attends the pain attack. The fever ends with the attack. If fever persists, it indicates its connection with inflammatory complication of cholelithiasis. This is confirmed by an increase in leucocytosis, ESR, and a sharp deterioration of the patient's general condition.

The patient may sometimes be obese, with xanthomatous plaques (cholesterol deposits) on the upper eyelids (less frequently on the other parts of the skin). The abdomen is distended; surface palpation reveals tension of the anterior abdominal wall, especially in the region of the right hypochondrium, and also excessive tenderness of the region. As pain is abated, the muscular tension subsides, and the tender edge of the liver can then be palpated. The gall bladder can sometimes be palpated as an oval or pear-shaped elastic body.

Tender points and sites of hypersethesia can sometimes be determined on the body according to Zakharyin and Head. These are as follows: (1) the region of the gall bladder (its projection on the skin); (2) eipgastrium; (3) pancreato-billary –cystic point; (4) shoulder zone;(5) point of the scapular angle; (6) paravertebral points to the right of the 8th to 11th thoracid vertebra; (7) phrenic nerve site (tender to pressure in the region between the anterior heads of the right sternocleidomasstoid muscle (phrenic symptom, or de Mussy-Georgievski symptoms).

Some laboratory and instrumental studies reveal signs of cholelithiasis in the full absence of its symptoms. Blood test shows an increaed cholesterol content. Duodenal probing (carried out in remission) can sometimes reveal fine stones (microliths) and a large quantity of cholesterol crystals. The most important diagnosistic technique in cholelithiasis is contrast roentegenography (cholecysto- or cholangiography and echography studies). These techniques help reveal stones in the gall bladder and the bile ducts.

Course and complications

The course of cholelithiasis is quite varied. Non-complicated cholelithiasis can namifest itself by only one attack of gall-stone colic. The attacks however are usually recurrent. They follow one another at short intervals or can occur once in 1 or 2 years, and even less frequently. Rare cases are known when the patient recovers spontaneously with the discharge of a small stone into the intestinal lumen. Long-standing cholelithiasis is usually attended by infection. The main disease is then aggravated by symptoms of cholecystitis or cholangitis.

Obstruction of the gall-bladder neck is a complication of the disease. It can cause hydrops. Obstruction is manifested by a most sever pain attack. Following several weeks, an enlarged gall bladder can be palpated. It is elastic and painless. In the absence the gall-bladder adhesion to the neighbourign organs (Due to paricholecytitis) the bladder can easily be displaced together with the liver durign deep respiration and by palpation.

In hydrops, the gall bladder is filled with a yellowish or colourless fluid (white bile) which is formed due to absorption of bile elemetns in the gall-bladder walls and effusion of serous exudate from the gall-bladder mucosda. If an infection joins, empyema of the gall bladder develops and the patient's condition is sharply deteroirated. The patient feels chills;; the body temperature is high; pain the the right hypochondrium develops again. Neutrophilic leucocytosis and increased ESR ar characteristic. If the gall-bladder entrance is fully obstructed by a stone, the bladder may become gradually affected by cirrhosis and its walls become sclerosed.

The common bile duct may be obstructed by a stone that passes from the gall bladder. As a rule, the stone is retained at the sphincter of the hepatopancreatic ampulla. Soon after the pain subsides, signs of obstructive jaundice develop. The common bile duct is obstructed completely when a mechanical closure (by a stone) combines with a spasm and inflammatory oedema of mucosa of the bile duct (cholangitis) that hinder bile outflow. The gall bladder does not usually increase despite congestion because its wall are affected by the attending inflammation and are no longer distended.

Bile outflows to the duodenum when stones move from the narrow to the wider part of the common bile duct (valve stones), or during transient relaxation of the gall-bladder walls. Jaundice intensity thus increases and decreases periodically; the colour of faeces changes accordingly.

Another complication of the disease is perforation of the gall bladder (less frequently of the common bile duct) with development of the outer or inner vesico-intestinal passage and sometimes bile peritonitis. Long presence of stones in the gall bladder can cause cancer, and prolonged obstruction of the common bile duct with bile congestion and infection of the bile ducts often provoke biliary (cholestatic) cirrhosis of the liver.

Treatment

Conservative treatment provides better outflow of bile and decreases the tendency to further formation of stones. The patient is recommended to lead a more active life, and prescribed frequent meals with restricted intake of cholesterol-containing foods, mineral water, and cholagogues. Various antispastic and pain removing preparations are prescribed (atropine, papaverin, warmth, etc.)

Surgical treatment of cholelithiasis is indicated in the presence of hydrops or empyema of the gall bladder, obturation of the common bile duct with obstructive jaunice, perforation of the gall bladder with development of fistulae or bile peritonitis, or in the presence of frequent attacks of gall-stone colics that fail to be removed by conservative treatment.

Prophylaxis consists in removal of metabolic disorders and cause of bile congestion. The patient is recommended regular meals, exercises, active mode of live, rational diet, and measures to prevent constipaton.

Cholecystitis

Cholecystitis is the inflammation of the gall bladder. The incidence of the disease is rather high; women are mostly affected.

Aetiology and pathogenesis

Various infections, autolytic affections of the gall-bladder, and helminthis invasions are important factors provoking cholecystitis. Virus aetiology of cholecystitis has been recently proved (Botkin's disease virus). Cholecystitis of toxic and allergic nature also occur. The aetiological role of infection in development of cholecystitis is confirmed by bacteriologicla studies of microbial flora of B bile obtained during operation or by duodenal probing. Infection may enter the gall bladder by enterogenic (from the intestine), haematogenic (From remote foci of infection such as affected btonsils, carious teeth, etc.) and lymphogenic routes. The aetiological importance of lamblia in the development of cholecystitis is disputable.

Bile congestion in the gall bladder predisposes to cholecystitis. The disease can be provoked by gall stones, dyskinesia of the bile ducts (under the effect of various psychoemotional factors, endocrine disorders, dysfunction of the vegetative nervous system, numerous nerve reflexes of the pathologically changed organs of the digestive system, etc.), anatomical properties of the gall bladder and bile ducts, ptosis of the internal organs, pregnancy, inactive mode of life, rare meals, habitual constipation, etc. Acute and chronic cholecystites are differentiated.

Acute Cholecystitis

Pathological anatomy

Acute catarrhal cholecytitis is characterzed by a mildly enlarged gall bladder containing serous or seropurulent exudate. The mucosa is swollen and plethoric. The inflammation affects the submucus layer as well (it is infiltrated by leucocytes). In purulent forms of cholecystitis, the lumen of the gall bladder is filled with purulent exudate; its walls are richly and diffusely infiltrated by leucocytes. The mucosa is oedematous, hyperaemic; erosion is frequent; deep ulcers or necrotic affection of the entire depthof the bladder wall occur in graver cass, gangrenous cholecystitis (gangrene of the gall bladder).

Clinical picture

Acute cholecytitis begins vigorously: sharp pain arises in the left hypochondrium which involves the entire upper abdomen and radiates into the right side of the chest, the neck, and sometimes into the hart. Pain may resemble biliary colic, but it is less intense. Pain continues for a few days; if not treated, it may continue for longer time. Pain is often attended by nausea and vomiting with a small amoutn of bile. Pain arises as a result of inflammation of the wall and serous coat of the gall bladder and distensionof the overlying peritoneum. The temperature rises to 38 and even 40°C and the patient feels chilly. Sometimes a mild jaundice develops due to inflammatory oedema of the mucosa of the common bile duct and obstructed bile outflow. The tongue is dry and white-coated. The abdomen is distended, the movements of the anterior wall ar limited, or the wall is not involved in the respiratory act at all.

Surface palpation first reveals local and then diffuse tension in the abdominal wall and sharp tenderness in the right hypochondriu. Acute cholecystitis is also characterized by some othe r symptoms: Zakharyin's symptom (sharp pain in the region of the gall bladder when it is tapped or pressed), Vasilenko's symptom (sharp pain in the region of the gall bladder when it is tapped over at the height of inspiration), Obrazttsov-Murphy symptom (sharp pain in the right hypochondrium when the examiner's hands press the gall bladder at the height of inspiration). Ortner's symptom (pain during tapping over the right costal arch by the edge of the hand). If inflammation extends onto the peritoneum overlying the gall bladder, Schetkin-Blumberg symptom ispositive. In this case, in the presence of gangrenous cholecystitis (gangrene of the gall bladder) and possible performation of the ball-bladder wall, a dangerous sign appears, i.e. the peritoneum friction sound at the point of its projection onto the abdominal wall. In moderate tension of the abdominal muscles it is sometimes possible (especially in purulent cholecystitis) to palpate an enlarged and very tender gall bladder. The liver does not usually increase, but its tender edge can sometimes be palpated. The de Mussy-Georgievsky symptom (tenderness at the point of the phrenic nerve, between the heads of the sternocleidomastoid muscle) can often be positive. Zones of skin hyperaesthesia (Zakharyin-Head symptom) can be found below the inferior angle of the right scapula and in the region of the 9th-11th interspaces. Leucocytosis is shifted to the left and the ESR increases.

Duodenal probing (that can only be done during abatement of the process) often fails to obtain B bile; or a cloudy whitish exudate poor in bilirubin can only be obtained. Bile samples contain much leucotyes, mucus, and cells of desquamated epithelium. The corresponding flora can be revealed in bile cultures.

Course

Patients with catarrhal cholecystitis recover comparatively soon. But the disease may convert into the chronic form. Acute purulent cholecystitis has a graver course, with signns of toxicosis, peritoneal irritation, high neutrophilic leucocytosis and considerably increased ESR. Signs of general toxicosis are more vivid in gangrene of the gall bladder, while in its perforation the symptoms of peritonitis join the picture.

Treatment

The patients must be admitted to hospital. Extirpation of the gall bladder is indicated in purulent and gangrenous forms of acute cholecystitis. Strict bed rest and abstention from taking food during the first two days following attack are prescried to patients with catarrhal cholecystitis. Later food should be given in small portions, 5-6 times a day, according to Pevnzer. Broad-spectrum antibiotics (oletetrin, 100 mg, 2-3 times a day intramuscularly, during 5-7 days) and spasmolytics (papaverine hydrochloride, 2 ml of a 2 per cent solution, 3 times a day subcutaneously) should be given.

Chronic Cholecytitits

Chronic cholecystitis may develop after actue cholecystitis but in most cases it develops gradually as an independent disease.

Pathological anatomy

The inflammatory process affects all layers of the gall-bladder wall in *chronic cholecystitis*. The bladder wall gradually scleroses, grows thicker, and calcium is deposited in the tissue. The gall bladder diminishes in size and adhers to the neighbouring organs. The adhesions deform the bladder to interfere with its normal function and to provide conditions for inflammation with periodic exacerbations.

Clinical picture

The patient complains of dull boring pain in the right hypochondrium which usually develops 1–3 hours after taking abundant (especially fat and roasted) food. The pain radiates upward to the region of the right shoulder, neck and the scapula. If cholecystitis concurs with *cholelithiasis*, sharp pain may arise (like in biliary colic). *Dyspeptic* signs are also present: bitter and metallic taste in the mouth,eructation, nausea, abdominal flatulence,and alternationof diarrhoea with constipation. The disease is sometimes not attended by pain except that the patient feels heaviness in the epigastrium or right hypochondrium, and dyspepsia develops. The temperature is often subfebrile.

The appearance of the patient and his nutrition are usually normal. Moderate obesity is sometimes observed. Examination of the abdomen can reveal its flatulence (either uniform or predominantly in the upper portions).

Surface palpation of the abdomen reveals sensitivity and sometimes pronounced tenderness in the region of gall-bladder projection. Muscular reistance of the abdominal wall is usually absent. De Mussy-Georgievsky, Ortner's, Obraztsov-Murphy, and Vasilenko's symptoms are positive. The liver is usually of normal size but in the presence of complications, such as hepatitis or cholangitis, the liver may be slightly enlarged with firm and tender (to palpation) edge. The gall bladder is impalpable.

The blood changes (during exacerbation) are characterized by moderate leucocytosis and mildly increased ESR.

Signs of *inflammation* (mucus, leucocytes, desquamated epithelium) can be found in B bile. If inflammation involves bile ducts (cholangitis), C bile containsn the same signs of inflammation. The vesical reflex (B bile) is sometimes impossible to obtain even by repeated probing. This indicates disordered contractility of the gall bladder which is typical of chronic cholecystitis. Bactriological studies of B bile reveal the character of microbial flora. Polarographic study of bile can reveal signs of inflammation.

Cholecystography shows change in the configuration of the gall bladder and the absence of its distinct contours. This indicates upset concentrating capacity of the gall-bladder mucosa. After taking a stimulating meal the gall bladder contracts insufficiently.

Course

The course of the disease is characterized by alternation of exacerbations and remissions. The disease can be exacerbated by abuse of fatty or fied foods, smoked meat and fish, condiments, alcoholic drinks, etc., by acute intestinal infections, and other factors. The process continues for many years and even decades. Cholecystitis is often complicated by inflammation of the bile ducts (*cholangitis*) or the pancreas (*pancreatitis*).

Treatment

Patients with exacerbated chronic *cholecystitis* must be treated in in-patient conditions (like in acute cholecystitis). In interparoxysmal period, treatment should be given 1–2 times a year in anti-relapse courses: periodic duodenal probing or giving cholagogues in 3–4 week courses (e.g. allochol per os, 1–2 tablets 3 times a day after meals, cholagogic species in the form of infusions, 10–20 : 200 ml, half-glass 3 times a day, 30 minutes before meals). Sanatorium and health-resort therapy is also indicated.

Prophylaxis

The disease and recurrent exacerbations should be prevented by taking measures to control bile congestion (exercises, walks and trips, regular and frequent meals with certain restriction) and treat focal infections.

MAJOR CLINICAL SYNDROMES

Exocrine Pancreatic Insufficiency

This symptom complex is characterized by disordered secretion by the pancreas of its juice containing the main digestive enzymes, such as trypsin, lipase, amylase and others (over 15 altogether), and also hydrocarbonates that ensure the optimum medium for the activity of thse enzymes. Exocrien insufficiency of the panceas can be primary (congenital) and secondary (acquired). Primary exocrien insufficiency of the pancreas can be due to its underdevelopment and mucoviscidosis (congenital systemic *cystofirosis* of the exocrine glands, e.g. the pancreas, bronchial, salivary or sweat glands, which is manifested by increased viscosity of their secretion due to high mucopolysaccharide content). The secondary exocrineinsufficiency of the pancreas arises in the presence of any disease attended by the affection of a considerable part of the pancreatic parenchyma and by obstruction of the secretion outflow.

If the pancreatic juice is delivered to the intenstine in deficient quantity (normally from 1.5 to 21 per 24 h) or if the amount of the main enzymes in the juice is deficient, normal digestion is upset and conditions for acclerated reproduction of microorganisms in the small intestine are created to cause dysbacteriosis which upsets digestion to a greater extent. Rumbling and sound of pouring liquid are heard in the abdomen; the patient suffers frommeteorism and characteristic pancreatogenic diarrhoea (polyfaecalia, yellowish faeces with fatty lustre). Coprological studies reveal steatorrhoea, creatorrhoea,and amylorrhoea. But early stages of pancreatic inufficiency cna proceed without marked intestinal dysfunction because of the high reserve potentialities of the pancreas.

Owing to the upset intestinal function, undigested food particles can be determined coprologically. Apart from the cavitary digestion the parietal intestinal digestion (affected mainly by the intestinal enzymes) and the absorption of the producs of enzymatic hydrolysis are also upset. The patient develops cachexia, signs of polyhypovitaminosis, symptoms of deficiency of the major microelements (iron, manganese, ions of calcium, sodium, potassium etc.) are observed. The function of many endocrine glands is upset secondarily. The patient complains of general weakness and decreaed work capacity. Hypoproteinaemic oedema can develop in patients with cachexia.

Pancreatic secretion is studied for the diagnosis of the syndrome of pancreatic exocrine insufficiency. Activity of trypsin, antitrypsin, lipase, and amylase in the blood serum, and also amylase in the urine is studied simultaneoulsy. Upset intestinal digestion, which is characteristic of pancreatic insufficiency, is determined in grave cases by the typical disorders in stools. Since in most diseases of the pancreas its endocrine apparatus is also involved (pancreatic islands) determination of blood sugar (with a fasting stomach), the level of glycaemia and glucose tolerance (single and double loads) is diagnostically important.

It should be remembered that the exocrine pancreatic insufficiency occurs in many diseases. The diagnosis of the main disease (by direct examination oand laboratory studies) is therefore of primary importance for further treatment of the patient.

Three stages of exocrine panceaatic insufficiency are differentiated; the first stage (initial, latent) become only manifest when the requirement for the digestive enzymes increases (overeating, especially intake of much fats); the second stage is a pronounced pancreatic insufficiency (frequtnt or permanent diarrhoea, steatorrhoea, creatorrhoes, amylorrhoea); and the third stage, dystrophy; it is characterized by considerable wasting (to cachexia) due to severe disturbances in the intestinal digestion and absoprtion, polyhypovitaminosis, and dystrophic changes in various organs and tissus.

Treatment

The following three principles are observed: (1) treatment of the main disease; (2) mechanically and chemically sparing diet: food must be easily hydrolysed by the enzymes and contain much protein, vitamins, and limited quantity of carbohydrates (depending on the pancreatic incretory function); (3) substitution enzyme therapy including preparations containing pancreatic enzymes: pancreatin, pansinorm, and others.

Prophylaxis

Timely detection, treatment and also prevention of disease of the pancreas (rational diet, abstention from alcohol).

SPECIAL PATHOLOGY

Pancreatitis

Pancreatitis is inflammation of the pancreas. Acute and chronic pancreatittis are differentiated.

Acute Pancreatitis

Aetiology and pathogenesis

Most often three exists connection between acute pancreatitis and inflammation of bile ducts, cholelithiasis in particular. This connection is explained by possible penetrtaion of bile (usually infected) into the pancreatic duct and activation of the enzymes (trypsin and lipase) of the pancreatic juice. This condition can arise when the common bile duct and the pancreatic duct have a common ampulla, e.g. in spams of the sphincter of the hepatopancreatic ampulla, obstruction of the ampulla by a stone, increased pressure in the duodenum (during coughing, vomiting, etc). Upset outflow of the pancreatic juice is obstruction of theduct by a stone, oedema of the duct mucosa, etc. is also important.

Among the other aetiological factors are alcoholism, poisoning withchemicals, such as lead, cobalt, phoshporus, arsenic, etc., and alimentary disorders (overeacting or inadequate nutrition), certain infectious diseases (e.g. epidemic parotitis, virus hepatitis), local circulatory disorders in the pancreas due to spasm of the vessels, embolism and thrombosis arising due to general changes in the vascular system.

Activation of proteolytic enzymes in the pancreas is important in the paghogenesis of pancreatitis irrespective of its aetiology. This cause enzymatic digestion (autolysis) of the pancreatic parenchyma with haemorrhages and fat necrosis. The developmental mechanism of pancreatitis depend also on secondary infection of the excretory ducts that develops by ascending haematogenic or (less frequently) lymphogenic routes.

Pathological anatomy

Inflammation, necrosis, and in later periods atrophy, fibrosis and calcification of the pancreas are revealed. Abscesses of variablae size are found in purulent inflammation of the pancreas, or diffuse melting of the pancreatic tissue develops with subsequent fibrosis of the panceas (in favourable outcome). Mild forms of pancratitis are only manifested by inflammatory oedema of the gland.

Clinical picture

Actute pancreatitis occurs mostly in women with disordered fat metabolism (aged 30 to 60). The disease usually begins by a sudden pain in the upper abdomen which arises after taking ample and a fat food or alcohol. In mild cases the pain is nor severe and is mainly localized in the epigastrium. It may also be girdling with radiation into the lumbar region, the left shoulder blade, and sometimes the retrosternal region. Grave cases (acute necrosis of the pancreas) are manifested by excruciating pain which ends in collapse and shock. Pain is attended by nausea, painful vomiting, salivation, constipation or, less frequently, diarrhoea.

Inspection of the patient reveals pallid and sometimes icteric skin and mucosa due to difficult bile outflow from the common bile duct. In grave cases, general cyanosis is possible, or cyanosis may be local, on separate parts of anterior abdominal wall or the lateral parts of the abdomen. Cyanosis is connected with pronounced toxicosis. The abdomen is often inflated. Surface palpation of the patient with the early initial stage of the disease reveala a soft and tender abdomen; the left part is more sensitive. Later, when peritonitis joins the process, the muscles become strained and symptoms of peritoneal irritation develop. Ascites can be revealed in acute haemorrhagic pancreatitis. The pancreas is usually impalpable. Cases with skin hyperaesthesia in the upper left quardant often occur.

The temperture is subfebrile, high in nectotic or purulent pancre-atitis, and subnormal in collapse.

The blood tests show neutrophilic leucocytosis with a shift to the left, lymphopenia, aneosinophilia, and increased ESR. Durign the very first hours of the disease, increased quantities of the pancreatic enzymes (diastase and lipase) are contained in the blood and urine. The blood and urine amylase content can nevertheless remain normal in necrotic pancreatitis (or amylase content may even be decreased). In these cases, the decreased blood calcium content and increased actiity of aspartate aminotransferase are of certain diagnositic importance. Attacks of tetany can develop in marked hypocalcaemia. Hyperglycaemia and glucosuria are not infrequent.

Course

Acute pancreatitis lasts several weeks and can end by complete recovery or actue disease may convert into chronic and relapsing pancreatitis. If pancreatitis is severe, the patient may die durign the initial period of the disease from a collapse and shock, or later from grave complications (cysts and abscesses of the pancreas).

Treatment

The patient must be taken to hospital. The conservation therapy includes: (1) control ofshock (intravenous drop infusion of 2–3 l of a 5 per cent glucose solution, transfusion ofblood or plasma); (2) physiological rest for the pancreas with abstenttionfromfood during 2–4days); (3) adminsitration of antienzymatic preparations (tasilol, etc.) for inactivation of proteolytic enzymes; (4) inhibiton of pancreatic secretion and removal ofpains (atropine sulphate, promedol, paranephric orparavertebral block); (5) prevention of secondary infection (prescription of antibiotics). Surgical treament is indicated is suppuration of the pancreas, in peritonitis, and haemorrhagic pancreonecrosis.

Chronic Pancreatitis

Chronic pancreatitis occurs mostly in women aged 30 to 70. It can develop after acute pancreatitis or directly as a chronic condition due to the same aetiological factors upon which the onset of actue pancreatitis depends. Chronic pancreatitis in men occurs mostly due to chronic alcoholism.

Pathological anatomy

Morpholical changesoccurring in pancreatitis are oedema of the pancras, small haemorrhages, necrosis and proliferation of connective tissue with gradual atrophy of cell elements. Reparation of the pancreas occurs simultaneously (areas of hyperplasia and formation of adenoma). Sclerosis develops in the interstitial tissues and in the parenchyma of the pancreas. Therefore the insular cells, which remain intact for a long time, are later atrophied adn sclerosed. At the early stage of the disease the pancreas is only mildly enlarged and dense; cicatrices, calcification, and obstruction of the ducts develop later. The pancreas diminishes in size and becomes dense and cartilaginous.

Clinical picture

Patient with chronic pancreatitis complain of pains that come in attack or are permanent. Pain usually occurs in the upper abdomen or in the epigastrium and radiates to the left shoulder, shoulder blade, neck, or to the left iliac bone; it can sometiems be girdlingand radiate from the spigastrium, along theleft costal edge, to the spine. Pain intensifies after taking fatty food. The patient complains of poor appetite, aversion to fats, regurgitation, nausea, vomiting, abdominal flatulence, diarrhoea (constipation in some cases), and loss of weight. The characteristic symptom of chronic pancreatitis is ample grey and fetid fatty faeces (steatorrhoea) which is connected with developing exocrine pancreatic insufficiency.

The skin and thesclea of the patient are sometimes icteric dueto compression of the common bile duct by an enlarged head ofthe pancreas. Deep palpation of the abdomen reveals tenderness in the region of the pancreas. The gland can sometimes be palpated in emaciated patients as a dense band. Zones of hypersensitivity of the skin can also be revealed.

Neutrophilic leucocytosis and increased ESER are observed in the blood in grave cases. The content of thepancreatic enzymesin the blood and urine during eacerbation increases, but remains normal or even decreases in the atrophic process. The enzyme content in the pancratic juice in patients with severe affections of the pancreas is decreased. Hyperglycaemia and glycosuria can occur in some cases. Coprological studies show signs of inadequate digstion of proteins and fats (steatorrhoea, creatorrhoea; etc.) which is connected with pancreatic hyposecretion.

X-ray examination of the duodenum in conditions of artificial hypotonia (duodenography) reveals dilation and deformation of the duodenal loop due to enlargement of the head of the pancreas. Diagnosis is confirmed by echography.

Course

The disease is usually protracted, with periodic remissions and exacerbations. But the prognosis is usually favourable in the absence of pronounced pancreatic dysfunction or complications such as diabetes mellitus, etc.

Treatment

Bed-rest, rational and sparing diet that does not stimulate pancreatic secretion but contains sufficient amount of proteins and veitamins should be recommended during exacerbations; antiobiotics and anti-enzyme preparations (trasilol and others) shold also be given. Pancreatin, pansinorm and other enzymes preparations should be given in substitution therapy.

Prophylaxis

This includes timely treatment of disease that might be aetiologically significantly in the origin of chronic pancreatitis (disease of the bile ducts, etc.) and control of alcoholism.

4

Urinary System

MAIN CLINICAL SYNDROMES

Renal Oedema

Oedema of renal aetiology is qute specific in most cases and can easily be differentiated form oedema of other origin, e.g. cardiac oedema, by the affection of loose connective tissue (the eyelids, the face) rather than of the lower extremities. Renal oedema can develop and resolve quickly. In pronounced cases, oedema isusually uniform over teh entire trunk and the extremities (anasarca). Not only the skin but also subcutaneous fat and the internal organs become oedematous. The liver usually becomes oedematous and enlarged, but in renal diseases the enlargement of the liver is usually proportional to enlargement of the other organs, and is never so pronounced as in cardiac oedema. Greater or lesser amount of fluid is accumulated in the serous cavities, e.g. in the pleural, abominal, and percardial cavities. Oedema can be revealed by palpation. It can also be confirmed by the McClure-Aldrich test: 0.2 ml of isotonic sodium chloride solution is injected into the skin on the median surface of a forearm and the time of disappearance of the resulting weal is noted. In a healthy subject, the weal is resolved within one hour. Inthe presence of a marked oedematous syndrome, the dynamics of oedema during treatment can be better assessed by repeating the test in several days with measurement of girths of the extremities and the abdomen at the samelevel, by determiningthe fluid level in the pleural and abdominal cavities, by weighting the patient, and also by determining daily diuresis and water balance of the body (the ratio of the taken and eliminated liquid during 24-hours period).

Oedema, like the general disorder inteh water-salt metabolism, arises due to various causes in renal diseases.

1. Diffuse increased permeability of teh capillary wall is important in development of oedema in many diseases of teh kidneys attended by the oedematous syndrome. Great importance in this process is attributred to auto-immune processes and increased hyaluroinidase activity of the blood serum, which as a rule, attends many diseases of the kidneys. Hyaluronidase internsifies depolymerization of hyaluronic complexes of mucopolysaccharides that form the intercellular substances(interendothelial "cement")and the basal membrane of the capillary wall. Porosity of the wall thus increases. The decreased blood serum content of calcium is also important because calcium compounds with protein (calcium proteinate) is a component part of the intercellular "cement"; change in teh blood pH (acidosis) is important as well. Because ofthe generalized increase in capillary permeability, not only water and the dissolved substances, butalso much protein pass from the blood to teh tissues. Depolymerization of mucopolysaccharides of the intercellular substances of tissues increases the quantity of molecules in the intercellular fluid and raises its colloidal-osmotic pressure.

It follows that the nephorotic sydrome is characterized not only by increased permeability of the capillary wall that facilitates fluid transport to the tissues, but also conditions are provided forfluid retention in the tissues, because the increased colloidal-osmotic pressure of teh intercellular fluid accounts for its hydrophilic property: the intercellular fluid easier absorbs water and gives it back with difficulty. The comparatively high proteing content in the oedema fluid (transdudate) explains the highe rdensity and lower mobility of oedema in the presence of deranged capillary permeability compared with oedema associated with hypoproteinaemia.

In the presence of increased capillary permeability, transdduate is accumulated in the subcutaneous fat and other highly vascularized tissue. Serous cavities usually contain low amounts of fluid. Disordered capillary permeability in the glomeruli causes proteinuria and promotes developmentof hypoproteinaemia. Oedema of this type occurs not only in diseases of the kidneys but in some other diseases as well, e.g. it can also be allergic or angioneurotic (Quincke's oedema), in cases with bee stining, etc.

2. Colloidal-osmotic (hypoproteinaemic) mechanism of oedema development is also important in the nephorotic syndrome. It is manifested in a decreased plasma oncotic pressure due to high proteinuria which usually occurs in such patients, and alsoinprotein passage through teh porous capillary walls into the tissues. Oedema of predominantly colloidal-osmotic origin obeys the laws of hydrostatics and tends to develop in the first instance in teh lower extremities inwalking patients and in the loinof bed-ridden patients. Hypoproteinaemic oedema usually occurs in cases where the blood protein content isless than 35-40 g/l (3.5–4 g/100 ml) and albumins are contained in the quantity below 10–15 g/l (1–1.5g/100 ml). Qualitative changes in the composition of the blood proteins are very important. Highly dispersed proteins (albumins) are mainly lost int eh urine in nephritis patients; the amount of globulins decreases to a lesser extent. Osmotic pressure is determined by the quantity of molecules contained in a unit volume of blood plasma rather than by their molecular weight. The loss of highly dispersed albumins, whose specific colloidal-osmotic pressure is about three times that of coarse-dispersion globulins, thereforre substantially decreases oncotic pressure of the blood.

Hypoproteinaemic oedema arises not only in the nephrotic syndrome; it can also develop in long stravation (huner oedema), deranged absorption in the small intensine (disordered absorption syndrome), cancer cachexia, and in some other diseases attended by a decreased protein content in the blood plasma.

3. Hypernatriaemic oedema (to be more exact, hypernatriahistic oedema) is explained by teh retention of the highly hydrophilic sodium ions in the blood and especially in the tissue. Administration of sodium chloride in large doses can thus cause this oedema. Hypernatriahistia attending diseases of the kidneys is an additional factor intensifying the effect of increased capillary permeability and hypoproteinaemia. Hormone factors, and in the first instance hypersecretion of aldosterone (the adrenal cortex hormone) and antidiuretic hormone (the posterior pituitry hormone), are vey important in the accumulation of the sodium ion in some diseases.

Any oedema, irrespective of its intensity, indicates upset osmoregulation in which the hormone link (aldosterone-antidiuretic hormone system) is the decisive one. This hormone system is mainly responsible for maintaining constant volume and ionic composition of the blood. As the volume of circulating blood decreases even insignificantly (which can occur in renal diseases when part of the liquid passes from teh blood to tissues due to increased porosity of the capillary wall or decreased oncotic pressure of the blood), the volume receptors, located mainly in teh walls of the right atrium and the common carotids, are stimulated. Protective mechanisms respond to this stimulation to maintain the intravascular volume. Aldosterone secretion by the adrenal cortex is intensified to increase sodium reabsorption in the wall of the renal tubules and its concentration in teh blood, and to promote its accumulation in tissus. According to some authors, the quanitty of aldosterone excreted in the urine during 24 hours increases in teh nephrotic oedema from 2–10 to 25–200 μg and more. Sodium excretion in the urine therby decreases considerably. Secondary hypersecretion of aldosterone that develops as a compensatory reaction, e.g. in oedema or a sudden loss of water from the body, is called secondary hyperaldosteronism as distinct from the primary hyperaldosteronism that occurs in tumours or hypertrophy of the adrenal cortex. Increased sodium reabsorption in the renal tubules is followed by increased reabsorption of water. High concentrtion of the sodium ions in blood (due to its intensified reabsorption in the renal tubules) stimulates osmoreceptors and intensifes secretion of the antidiuretic hormone by the pituitary gland, which in turn intensifies the facultative reabsorption of water in distal tubules still more. If the primary cause of oedema (increased capillary permeability, decreased oncotic pressure of plasma) is still active, fluid is not retained in the blood vessels and continues its passage from the blood to the tissues to intensify oedema.

4. Oedema can occur in acute anuria of the kidneys in acute poisoning (e.g.with corrosive sublimate), hypovolaemia reduction of blood circulation in the kidneys (profuse blood loss, shock), and also in the terminal stage of certain chronic renal diseases (retnetion oedema). But decreased glomerular filtration becomes only important in the presence of other forerunners of oedema rather than an independent factor. For example, in severe renal insufficiency attended by pronounced filtration disturbances, oedema is often absent or even resolved, if any.

It should also be noted that none of the above mechanism of renal oedema develops independently but becomes only a dominating factor in this or that case.

Nephrotic Syndrone

The nephrotic syndrome (symptom complex) is characterized by pronounced proteinuria, hypoproteinaemia (mainly) due to hypoalbuminaemia), hperlipidaemia (hypercholesterolaemia), and oedema.

The *nephrotic syndrome* occurs in chronic *glomerulonephritis*, amyloidosis, malaria, sepsis, tuberculosis, collagenosis, diabetes mellitus, and certain other diseases. Less frequently the cause of the nephrotic syndrome cannot be established immediately, but in most cases a detailed analysis of anamnestic data and a thorough examination of the patient reveal chronic glomerulonephritis. These forms of the nephrotic syndrome occur mostly in children. Cases where the cause of renal systrophy is unclear are identified as lipid nephrosis.

It is believed that the nephrotic syndrome is caused by metabolic disorders, mostly upset fat and protein metabolism, with subsequent derangement of trophics and capillary permeability in the glomeruli. Protein and lipids contained in large quantity in primary urine of these patients infiltrate the tubular wall to cause drastic dystrophy in the epithelial cells. The auto-immune mechanism is of great significance for teh development of the chronic nephrotic syndrome. It has been proved by animal experiments: small doses of nephrotoxic serum provoke nephritis in rabbits; a picture characteristic of the nephrotic syndrome develops on administration of large doses.

Pathological anatomy

The following morphological signs of the nephrotic syndrome develop, in addition to the changes characteristic for the main disease. The kidneys are enlarged ("large white kidneys") and their capsule is easily removed. Histological studies reveal dystrophic changes in the epithelium of the tubules, especially of the convoluted tubules. Lipid deposits can be found in teh basal parts of teh epithelial cells. The glomeruli are affected by dystrophy; especially specific are changes in podocytes and endothelial cells with whichthe disordered permeability of teh glomerular membrane is associated.

Clinical picture

The main, and often the only complaint of patients is persistent *oedema*. It is especially pronounced on the face which becomes swollen andpallid, the eyelids are only a narrow slit, and the patient opens his eyes in the morning with difficulty. The legs, the loin, the skin of the abdomen and the hands are also affected by oedema. The oedema is mobile when the skin is pressed by the finger, a depression remains in it which soon disappears. Fluid is accumulated also in the internal organs and the serous cavities. In typical cases the arterial pessure remains unchanged or even decreased.

As the oedema progresses, diursis usually decreases and the patitn eoften eliminates only 250- 400 ml or urine a day. The specific gravity of the urien is high (1.030– 1.040) and it contains much protein, to 10–20 g/l and more. Cases were reported where teh urine contained 24 g/l of protein. Fine dispersed moleculesof albumins prevalil among protein. It is believed that the increased filtration of the plasma protein through the glomerular capillary wall and also disordered reabsorption of protein molecules by the affected tubular epithelium are important in teh aetiology of proteinuria in the nephrotic syndrome. Great quantity of hyaline, granular and waxy casts, and cells of renal epithelium are found in the urinary sediment. The presence of leucocyes and erythrocytes in teh urinary sediment is not characteristic for the nephrotic syndrome. Doubly refracting cholesterol crystals are usually found, which are, as a rule, absent in renal diseases proceeding without the nephrotic syndrome.

A long-standing and persistent proteinuria causes protein depletion of the body and a stable reduction of its content in the blood plasma (1.5 and even 2 times). The albumins become especially deficient, and the albumin to globulin ratio, which is normally 1.2–2.0, decreases significantly. he content of a2-globulins and also g-globulins, slightly increases. *Proteinuria* and *hypoproteinaemia* (especially hypoalbuminaemia) largely account for oedema that develops in the nephrotic syndrome. It has been established that the loss of protein in the urine is aggravated by the renal catabolism of plasma proteins (proteolysis of part of serum protein durign its reabsorption in the tubules) and also, probably, by the increased protein loss through teh alimentary tract.

Among the constatn symptoms are pronounced *hyperlipidaemia*, increased blood serum concentration of cholesterol (to 13–15 mmol/l, i.e. 2 or 3 times as great), phsopholipids, and neutral fat. Thee changes are probably secondary to upset protein metabolism and hypotroteinaemia. Laboratory studies reveal three characteristic signs of the nephrotic syndrome: proteinuria, hypotrpteinaemia, and upset lipid metabolism (hypercholesterolaemia).

The blood clearing function of the kidneys is not substantially affected in the *nephrotic syndrome*, and *azotaemia* does not develop for a long time. The main functional renal tests remain normal for a long long time, but the tubular secretion can decrease. Biopsy of the kidneys supplies valuable information concerning the nature of the nephrotic syndrome is chronic renal diseases.

Course

If the main disease does not prgress, thenephrotic syndrome lasts for years. Oedema and the urinary syndrome intensify at times usually when provoked by an attending infection. Patients with teh nephrotic syndrome are sensitive to coccal infection. They often develop recurrent pneumonia and crysipeloid inflammation of the skin. These patients usually died before antibiotics were discovered. Vascular thrombosis is likely to occur in patients with the nephrotic syndrome. The prognosis depends mostly on the main disease and the attending infections.

Treatment

The main disease should be treated. In the presence of pronounced hypoproteinaemia, the patient is prescribed a diet rich in proteins (2–2.5 g/kg body weight without reference to oedema) and poor in sodium chloride. Plasma or concentrated human albumin is given intravenously. Corticosteroids (prednisolone) and immunodepressants (imurane, etc.) are prescribed. If oedema is pronounced, the patient is given diuretics; furantril (furocemid), 0.04 g per os evey other or third day, in combination with verosperon (0.075–0.15 g/day per os) to remove oedema. Sanatorium and health-resort therapy in dry climate (Central Asia) is recommended durign relative remissions.

Renal Hypertension

Renal arterial hypertension is a symptomoatic hypertension caused by the affection of the kidneys or renal vessels and upset renal mechanism of arterial pressure regulation. Amon gall cass of arterial hypertension, renal hypertension makes about 10–15 per cent.

Many diseases of the kidneys, in the first instance acute and chornoc glomerulonephritis, pyelonephritis, nephrosclerosis and various affectiosn of the renal blood vessels are attended by elevated arterial pressure. This is underlain by the important role that kidneys play in the regulation of arterial pressure. The juxtaglomerular apparatus of teh kidneys, which is an accumulation of special cells at the vascular pole of teh glomerulus at the point where the artery nears the proximal end of the distal convoluted tubule, produces renin in the presence of ischaemia of the renal parenchyma. Renin acts on the liver-produced hypertensiongen, which is fraction of a2-globulins of plasma, to convert it into angiotensiongen. The latter is converted enzymatically into agiotension (hypertensin). Angitensin stimulates hypersecretion of aldosterone, stenosis of arterioles, and increases arterial pressure. The current literature contains reports on teh devleopment of renal hypertension which is associated with renal hyposecretion of special hypotensive substances (e.g. prostaglandins).

It should be noted that due to diffuse spasm of arterioles, essential hypertension also provides conditions for atherosclerosis of the arteries, disordered blood supply to various organs, the kidneys included, and hence for the incerased secretion or renin . Therefore, at a certain stage of essential hypertension, further elevation of pressure in the arteries depends largely on the renal mechanism. As distinct from hypertension of other genesis, renal hypertension often tends to run an especially rapid and malignant course (in about 20 per cent of cases.)

Like any other hypertension, renal hypertension is manifested by some annoying subjective symptoms, such as headache, dizziness,and noise in the ears. Headache is especially serve in marked elevation of the arterial pressure. It is often attended by vomitting, and paraesthesia. As a rule, the work capacity is impaired and sleep deranged. The increased arterial pressure can be determined by feeling the pulse which appears to be tense. More accurately arterial pressure shoudl be measured instrumentally. Both systolic and diastolic pressure shoudl be measured; the latter pressure is sometimes very high. The second sound is usually accenturted over the aorta.

High and persistent hypertension affects the heart. First, the left ventricular muscle is hypertrophied due to constant overload. This can be determined by the increased apex beat, specific changes detectable by X-rays (rounded heart apex) and electrocardiographically (deviation of the heart's electrical axis to the left, a certain increase in the R1 wave, and later descent of the S-T1 segment below the zero line, and the negative or two phase T1,2 wave).

At later stages, dystrophic changes occur in the myocardium because its vascularization lags behind the growth of the muscle weight to account for the deficient blood supply; next, cardiosclerosis develops. At the same time, atherosclerosis of the coronary vessels may develop due to upset lipid metabolism, which is characteristic for arterial hypertension adn many other renal diseases attended by the nephrotic syndrome. The coronary disease impairs blood supply to the myocardium to an even greater extent. Heart pain, like that of angina pectoris often occus. Further progress of renal diseases can provoke circulatory insufficiency.

Certain acute diseases of the kidneys attended by a rapid and pronounced elevationof the arterial pressure, mainly acute glomerulonephritis, are attended by the condition at which the left ventricle isnot hypertrophied enough toe compensate for the markedly increased load. Acute ventricular failur can therefore develop. It is manifested by attaks ofcardiac asthma and even by a lung oedema.

Renal arterial hypertension involves specific changes in the fundus oculi (renal retinopathy). But these changes, attending kidney diseases, are often due not only to the spasma of arteries and arterioles of teh retina but also, in a certain measure, to the upset permeability of retinal capillareis; they also occur during final stages of chronic diseases of the kidneys that end withnephrosclerosis, uraemic toxicosis,and haemorrhagic syndrome.

Certain stenosis of arteries and arterioles of the retina, tortuosity of fine veins of the yellow spot, ant flatness of the veins at their crossing with the arteries, withsmall ampoule-like dilation before teh crossing (Hannn-Salus symptom), are observed during the first period (first degree). The patient does not complain of deranged vision; the changes are only transient and functional. Further, due to a continuing spasm and hyalinosis of the arteriolar walls, their lumen narrows, larger arteries become stenosed and tortuous, the veins are compressed by the crossing arteries, and their ampoule-like dilatations before the crossing point become more pronounced (second degree). In the final stage, the arteries and arterioles are highly tortuous and markedly affected by spams; they resemble silver wires; small veins of the retina are consolidated and sclerosed, they bend at teh point of crossing with the arteries and are impressed into the retina to simulate a rupture (the third degree). Greyish-white and yellow foci of oedema and retinal dystrophy and haemorrhages are revealed; thepapilla of the optic nerve isoedematous. Foci of dystrophy radiating in the area of the yellow spot often have a stellar shape. Vision is deranged due to dystrophic changes in teh fundus oculi and haemorrhages.

It should be noted that the fundus oculi is a very convenient site where fine vessels can be visualized, as well as changes occurring in them and the surrounding tissues (in the retina) during the course of the disease. The study of the fundus oculi is therefore very informative for the diagnosis and prognosis of a renal disease. In certain cases, the patient consults first the ophthalmologist for his impaired vision, because renal disease can for a long time develop without vivid symptoms. The specific changes in the fundus oculi help in such cases suggest renal pathology (which is later confirmed by the appropriate studies).

Finally, disorders in cerebral circulation with paralysis, deranged sensitivity, dysfunction of the pelvic organs, and also myocardial infarction can develop as a result of arterial hypertension and atherosclerosis.

It follows therefore that in certain kidney diseases, the renal hypertension syndrome can be primary significance in teh clinical picture of the disease and can be decisive for its course and outcome.

Renal Eclampsia

Eclampsia usually develops in acute diffuse glomerulonehritis, but can arise in aggravated chronic glomerulonephritis and nephropathy of pregnancy. The pathogenesis of eclampsia is largely underlain by increased intracranial pressure, oedema of the cerebral tissue and cerebral angiospams. Eclampsia in all these conditions usually arises in pronounced oedema and increased arterial pressure. Attacks of the disease are provoked by salted food and excess liquid.

The first signs of approaching eclampsia are often unusual somnolence and flaccidity. These are followed by severe headache, vomiting, temporary blindness (amaurosis), aphasia, transient paralysis, mental confusion, and a raid rise in the arterial pressure. Convulsions develop unexpectedly, sometimes after uttering a cry, or after a noisy deep inhalation. The convulsions are first strong tonic spasms, which are followed (in 0.5–1.5 minutes) by strong clonici contractions. Less frequently only twitching of some muscles is observed. The face becomes cyanotic, the neck veins swell, the eyds turn aside or roll up, the tongue is bitten, and foam emerges fromthe mouth. The pupils are dilated and do not respond to light; the eyeballs ar firm. The pulse is tense, slow, the arterial pressure increase. The body temperatures rises in frequent attacks. Involuntary defaecation and urination often occur.

Attacks of renal eclampsia usually last for a few minutes, rarely for longer time. Eclampsia occurs in some cases as a series of two or three attacks which follow one another. The patient then calms down to stupor, deep sopor or coma; consciousness is then regained. After revovery from the state of stupor the patient sometimes remains inamaurosis (blindness of the central origin) and aphasia (mutism).

This is the classical picture of an eclampsia attack. But it should be remembered that attacks of eclampsia may also be atypical; they may occur without loss of conciousness or occur in an obliterated form, as a transient aphasia, amaurosis, and mild muscular twitching.

Renal eclampsia should be differentiated from convulsions of other origin. Convulsions in eclampsia are the same as in epilepsy (a congenital or post-traumatic nervous disease). But oedema or other signs of renal insufficiently are absent in epilepsy; attacksof convulsion usually occur in the course of many years. Convulsions develop also in uraemic coma, but the patient has a typical *anamnesisin*, this case (chronic renal insufficiency) signs of *uraemic toxicosis*, slow (in the course of several days) development of the convulsive state; the character of convulsions is different as well: convulsions develop as slight fibrillary twitchings.

Treatment

An attack of *renal eclampsis* can be removed immediately by suboccipital or cerebrospinal puncture with extraction of small portion of the cerebrospoinal fluid: the intracranial pressure devreases and the patient regains consciousness. The extraordinary efficacy of cerebrospinal puncture proves the importance of increased intracranial pressure for the pathogenesis of attacks of renal eclampsia. *Phlebotomy* and intravenous injection of magnesium sulphate (10 ml of a 25 per cent solution) remove attacks of eclampsia, effectively decrease the arterial pressure and lessen cerebral oedema.

Renal Failure

Renal failure is toxicosis of the body caused by renal dysfunction (self-posioning). *Uraemia* is a severe form of renal failure. Renal failure and uraemia occur in acute and chronic cases. Acute uraemia develops in poisoning with nephrotoxic subsar es (compounds of mercury and lead, carbon tetrachloride, barbitur s, etc.), in transfusion of incompatible blood, profuse haemolysis, a n shock. *Chronic uraemia* develops in teh final stage of many chronic renal diseases, such as chronci glomerulonephritis, pyelonephritis, amyloidosis, affections of the renal vessels, tumours of the kidneys, etc.

Pathogenesis

It consists in profound homeostatic disorders. It has been established that products of protein decomposition are accumulated in the blood of patients with uraemia. These are nitrogenous slags, such as urea, uric acid, creatinine, and other guanidines. The content of indican, phenol and other aromatic substances that are formed in theintestine and pass into the blood through the intestinal wall (normally, these substances are eliminated from the blood by the kidneys) increases. Various compounds of sulphur, phosphorus, magnesium, and other substanes are accumulated; the ionic equilibrium is upset. *Acidosis* develops as a result of the accumulation of aciid products and disordered production by the kidneys of ammonia that neutralizes the acids. Uraemia is attended by a grave affection of the liver and metabolic disorders.

Acute renal failure and acute uraemia develop due mainly to shock and the accompanying circulatory disorder (mostly in the kidneys). *Anoxia* develops to cause dystrophic changes in the renal glomeruli and tubules. In other cases, when acute renal failure is due to poisoning or a grave infectious disease, its pathogenesis is largely determined by the direct action of poisons and toxins on the rennal parenchyma. In both cases glomerular filtration is deranged, diuresis decreases and oliguria develops; in severe cases anuria may occur. Salts of potassium, sodium, phosphorus,nitrous products and some other substances are retained in the body.

Acute renal failure rapidly develops and the patient's condition becomes grave: vomiting, mental confusion, deranged respiration and upset heart activity are observed. The glomeruli are affected by ischaemia to raise the arterial pressure; oedema develops in anuria. The patient may die unless *anuria* and *azotaemia* are removed during the first few days. If the course of the disease is benign, diuresis increases but the concentrating capacity of the kidneys remains impaired for some time; the renal function gradually normalizes and the patient recovers.

Clinical picture

Acute renal failure varies slightly, depending on the character of the main disease. In many cass it proceeds with some general symptoms which make a syndrome. Four stages of acute renal failure are distinguished: (1) Initial stage lasting from several hours to 6–7 days; its clinical picture is characterized by the main symptoms of the disease (*traumatic* or *transfusion* shock, severe infectious disease, poisoning, etc.); (2) *oligoanuric* stage characterized by changes in diuresis (to complete anuria), uraemic toxicosis, and water-electrolyte disorders. Proteinuria, cylindruria, and erythrocyturia are revealed on examination. The oligoanuric stage can end with death of the patient or his revovery. In the latter case, diuresis suddenly or gradually increases (the third or polyuric stage). The specific gravity of the urine is low, the concentration of residual products of protein metabolism in the blood decreases, water-electrolyte balance is restored and the pathological changes in the urine disappear. The fourth stage, recovery, begins with normalization of diuresis, it lasts from 3 to 12 months.

Development of chronic renal failure is determined by the progressive affection of the kidney parenchyma. The latent period of chronic renal failure, when renal dysfunction has no clinical symptoms and can only be revealed by special laboratory methods, and the manifested period, characterized by the marked clinical picture of uraemia, are distinguished.

The latent period can only be revealed by special tests carried out for concentrating capacity of the kidneys and cold food and by the Zimnitsky test. The patient's urine is usualy to low specific gravity (below 1.017). Variations in specific gravity are only insignificant (isohyposthenuria). The clearance tests reveal disordered reabsorption in the renal tubules and in glomerular filtration. Mild renal dysfunction can be revealed by radioisotope nephrography. It is belived that the first signs of renal failure in patients with chronic renal diseases only appear when the functioning parenchyma diminishes to at least one fourthof its normal size.

Progressive renal failure is attended by changes in the circadian variations in urination: isuria or nycturia are observed. The concentration and dilution tests reveal significant disorders inthe concentrting capacity of the kidneys, pronounced isohypothenuria (the specific gravity of all urien specimens varies from 1.009 to 1.011, i.e. approaches the specific gravity of plasma ultrafiltrate, the"primary" urine). More pronounced disorders in reabsorption and glomerular filtration are determined by teh clearance tests and nephrography. Concentration of nitrogenous substances in the blood gradually increases. Residual nitrogen increases several times (its normalcontent is 14.2–28.5 mmol/l or 20–40 mg/100 ml). Laboratory studies reveal increased concentration in the blood of various products of protein decomposition; urea (3.23–6.46 mmol/l in norm and 10-15 and more tiems higher in renal failure), creatinine (0.088–0.176 μmol/l in norm and 1–1.3 mmol/l in renal failure), indican (0.68–5.44 μmol/l) in norm). It should be noted that the increased blood indican content is often the first and the most reliable sign of chronic renal failure, because its blood content does not depend on the protein concentration of food and because it is not accumulated in tissues.

Moderately increased concentration of products of nitrogenous decomposition in the blood (*azotaemia*) may have no effect on the subjective conditon of the patient for a certain period of time. But later some external changes become manifest and they can be used for the diagnosis of uraemia. Certin symptoms of uraemia depend on partial compensation of renal failure by a more active involvement of the skin, mucosa, and the digestive glands in the excretory processes. Decomposition of the urea (excreted by the mucosa of the air ways and the mouth) to ammonia by teh bacteria accounts for the specific uaremic breath. In serious cases, the uraemic breath can be felt by the physician as he approaches the patient's bed. It is believed that the uraemic breath can be felt when the concentration of residual nitrogen in the blood exceeds 70 mmol/l (about 100 mg/100 ml).

The nitrogenous substances, and in the first instance urea, are liberated by the gastric mucosa and decomposed to form ammonia salts. These salts irritte the mucosa of the stomach and the intestine to stimulate nausea, vomiting (*uraemic gastritis*), and diarrhoea (*uraemic colitis*). Irritation of the respiratory mucosa causes laryngitis, tracheititis, and bronchitis. Severe *stomatogingivitis* develop. The mucosa becomes affected by ulcers and necrosis. Urea crystals (as a white powder) can sometimes be seen on the patient's skin. This is especially noticeble at the orifices of the sweat glands (at the base of hairs). Strong itching develops and the patients scratches his skin. Poisons accumulated in the blood are also liberated by the serous membranes. Uraemic pericarditis is especially characteristic. It can be revealed by auscult-ating the heart usign a stehoscope: the specific coarse pericardial friction can be heard. This friction appears in the terminal period and is a sign of approaching death.

Memory and sleep become deranged due to general poisoning; weakness, dull headache, somnolence, apathy and deranged vision are characteristic. Examination of the fundus oculi reveals narrowed arteries and dilated veins, oedema of the papilla of the optic nerve, and whitish local foci (*retinopathy*). Development of retinopathy is explained by trophic disturbances due to the vascular spasm of the fundus oculi vessels and uraemic toxicosis which intensifies these changes. The pupils are usually narrowed.

Metabolic disorders are pronounced: the patient develops cachexia; the liver and bone marrow functions are affected by dystrophy; toxic uraemic anaemia develops which is usually attended by leucocytosis and *thrombocytopenia.* The tendency to haemorrhages develops due to a decreased blood platelet count, disorders in the blood coagulating system and increased capillary permeability (as a result of toxicosis). Haemorrhages of the gastro-intestinal tract, urinary tract, uterus, and the nose may develop. Skin haemorrhages also occur. The body temperature slightly decreases.

Later, *toxicosis* increases, the patient's consciousnes becomes dimmened,and uraemic coma develops. Periods of stupor alternate with periods of excitation, *hallucinations*, and noisy slow breathing with very deep inspirations; respirtion with alternating periods of*hyperpnoea* and *apnoea* occurs less frequently. At the terminal stage the patient isin a deep coma; muscular twitchings occur at times and the patient dies.

There is no universally accepted classification of chronic renal failure at the present time.Three stages are usually differentiated; (1) the initial stage with insignificantly increased residual nitrogen and creatinine and moderately decreased glomerular filtration; (2) pronounced stages (IIA and IIB) with marked azotaemia and electrolyte disorders and (3) terminal stage with a pronoucned clinical picture of uraemia.

Main principles of treatment

Patients with acute renal failure should be immediately taken to hospital. Treatment should be begun as early as possible and aimed at eliminating the main aetiological factors (removal of nephrotoxic substanes or their detoxication, giving large doses of plasma or blood substitutes for hypovolaemic shock). The electrolyte equilibrium should be correctred simultaneously. Mannitol or furocemid should be given intravenously in the initial functional stage of acute renal failure to restore diuresis. Haemodialysis (artificial kidney) or peritoneal dialysis are necessary in grave cases.

The protein content of the diet is controlled inpatients with chronic renal failure. Therapeutic action on the aetiological factor and pathogenic mechnisms should be attempted (antibacterial therapy in pyelonephritis, immunodepressants in chronic glomerulonephritis, etc.). The water-electrolyte equilibrium and acid-base balance are corrected simultaneously. The necessary symptomatic therapy with hypotensive, cardiovascular and other preparations should be carried out. In severe cases, permanent chronic haemodialysis is carried out or the kidneys ar transplatned to prolong the patients's life.

SPECIAL PATHOLOGY

Disesase of the urinary system come among the first in the list of diseases the most frequently occurring. In some cass the kidneys may be the locus of primary pathology. These are comparatively rare congenital abnormalities and genetic nephropathies. Inflammatory affections of the renal parenchyma, such as diffuse glomerulonephritis of infectious allergic nature, pyelonephritis arising in inflammatory processes of teh urinary tract (cytitis, pyelitis) due to extension of inflammation onto therenal parenchyma, and focal nephritis in sepsis occur more commonly. The group of metabolic-dystrophic diseases of the kidneys includes their acute affections in vrious possonings with nephrotoxic substances, in shock, and also chronic diseases such as amyloidosis and, to a certain extent, nephrolithaiasis. The kidneys are oftne the site of tumours (especially frequently hypernephroid cancer or hypernephroma). Affections of the renal vessels are not infrequent. Traumatic affectiosn of the kidneys often occur in surgery and treatment of urinary diseases.

The other group includes affections of the kidneys that are secondary to diseases of teh other organs and systems: essential hypertension, atherosclerosis, diaetes mellitus, gout, collagenosis, general infections, etc.

In many untreated or improperly treated cases, primary or secondary affections of the kidneys cause dystrphic changes in nephrons, development of connective tissue with its subsequent cicatrization and cirrhosis of the kidneys (nephrosclerosis) which is manifested clinically by signs of progressive renal failure.

Diffuse Glomerulonephritis

Diffuse glomerulonephritis is the general infectious allergic disease with predominatn affection of the glomerular vessels. Acute and chronic glomerulonephritis are distinguished.

Acute Glomerulonephritis

Aetiology

Acute diffuse glomerulonephritis usually develops after acture infectious diseases, such as tonsillitis, scarlet fever, acute respiratory diseases, pneumonia, and otitis. Especially important are diseases caused by group A haemolytic sterptococcus, most freqeuntly of type XII> But nephritis can arise also after infectious diseases caused by other bacteria, e.g. pneumococci or staphylococci. Acute nephritis sometimes develops following overcooling, especially in damp weather. Cases were reported of acute nephritis developing after vaccination.

Pathogenesis

Acute nephritis typically arises not during an infectious disease but only following a period of time, usually 2–3 weeks later. Attempts to isolate the streptococcus from the kidney tissue end in failure. Thus, the onset of acute nephritis usually coincides with the period when antibodies to streptococcus are produced. This indicates that acute nephritis is not simply an infectious disease but an infectious allergic disease.

It is suggested that bacterial antigens, that get into the blood durign infection, injure the kidney tissue, whose affected proteins act as an antigen to stimulate the production of the corresponding antibodies inteh reticuloendothelial system. The antigen-antibody complexes are fixed in the endothelial and epithelial cells of the renal glomeruli adn also in the basal membrane of the glomerular capillaries to cause their injury. Both kidneys are always involved in acute diffuse glomerulonephritis and all glomeruli are equally affected. This distinguishes the affection from focal nephritis and confirms its allergic nature.

It is necessary to note that both the glomerular capillaries and vessels of the other organs and tissues are affected in acute glomerulonephritis. Nephritis is thus the general vascular disease. Cases have been describe when in the presence of a marked clinical picture of the disease (oedema, hypertension), the urinary symptoms were insignificant or absent. But as a rule, the glomerular apparatus of the kidneys is affected in acute nephritis which is explained by the specific character of their function as the excretory organ.

Pathological anatomy

The kidneys of those who died from acute nephritis are of normal size or slighly enlrged; their colour is bron or greyish-brown. Malpighi corpuscles, in the form of small tubercles, can be seen on secretion. Microscopy of the renal tissue in the initial stage of the disease reveals enlared and hyperaemic glomeruli; during further progress of the disease, microscopy reveals ischaemia of the glomeruli due to spasm of the capillary loops, fibrinoid swelling of the capillary walls, proliferation of their endothelium, accumulation of coagulated proteinous exudate in the space between the capillary loops and the glomerular capsule, blood stasis, thrombosis of the capillary loops, and haemorrhaes. Pathological chanes occur in both kidneys in all cases. Epithelium of the renal tubules is affected less markedly. Intra-and-extra capillary glomerulonephritis is mostly differentiated; depending on the character of the inflammation, glomerulonephritis may be exudative (serous, fibrinous, and haemorrhagic) and productive.

At later stages of the disease, inflammatory phenomena subside in renal tissue, proliferation of endothelium in glomerular loops decreases, and patency of the capillaries is restored.

Clinical picture

The clinical picture of acute glomerulonephritis is quite specific and is determined by teh main three syndromes: oedema, arterial hypretension, and changes in the urine (haematuria and proteinuria). The patients would usually complain of oedema, which arise first on the face, under the eyes, and then extend onto the entire body and the extremities. Development of oedema is explained by disordered capillary permeability and aldosterone hypersecretion by teh adrenal cortex. Headache and heaviness in the head are frequent symptoms. They are explained by increaed arterial pessure and, in some cases, intracranial pressure. Vision can be deranged due to spasm of the retinal vessels and haemorrhages into the retina. Many patients complain of general fatigue and reduced work capacity.

In the presence of a pronounced oedema and massive plerual effusion, and when the heart muscle is overloaded due to markedly increased arterial pressure, patients with acute, hephritis suffer from severe dyspnoea, sometimes with attacks of asphyxia (like in cardiac asthma).

The patient with acute nephritis would often complain of dull lumbar pain. The gravity of the disease depends on the degree of oliguria. The diuresis decreases while the patitne may have frequent tenesmus. Complate anuria occurs in some cases. If haematuria is marked, the urine looks like meat wastes.

Inspection of the patient reveals his specific appearance; pallid skin, oedematous face, swollen eyelids, and oedema of the trunk. Some patients assume the forced semireclining or sitting position because of pronounced dyspnoea. Renal eclampsia occurs in grave cases. The onset of an eclampsia attack is hearlded by increasing arterial pressure and a severe headache.

The extent and the character of oedema can be established by palpation. The pulse of the patient should also be felt. Acute nephritis is characterized by a tense pulse which is sometimes slow. The apex beat is somewhat shifted to the left and increased due to myocardial hypertrophy which soon develops in the presence of arterial hypertension.

Percussion of the chest in the presence of genealized oedema reveals free fluid in the plerual cavity (transduate) and congestion in the lung root region (dulled tympany). The left border of the heart extends beyond the correspondign midclavicular line.

Normal or harsh respiration is heard by ausculation. In the perence of pronounced congesion, dry and moist congestive relaes are head. Ausculation of the heart reveals bradycardia (due to the reflex transmitted in increased pressure from the aorta onto the vagus nerve through n. depressor).

The first sound is sometimes decreased at the heart apex. If the heart muscle is much overloaded, the gallop rhythm is heard. The second sound is usually accentuated over teh aorta due to increased arterial pressure.

X-ray studies of the chest confirm the presence of pleural effusion and congestion in the lung roots. Dilation and hypertrophy of the left ventricle are clearly determined (the heart apex is rounded).

Sphygmomanometry is of great help in establishing a diagnosis. It reveals one of the main symptoms of acute nephritis, i.e. arterial hypertension. Systolic pressure increases to 200-220 mm Hg, but in somen cases it is not so high. Diastolic pressure increases to 100–160mm Hg almost in all cases.

Electrocardiography reveals signs of hypertrophy and overload of the left-ventricular myocardium. The amplitude of ECG waves decreases in pronounced oedema of the trunk.

Changes in the urine are characteristic of acute nephritis. During development of oedema, diuresis usually decreases to oliguria. The urine of patients with acute nephritis usually conains much protein and erythrocyes due to the increased permeability of the renal capillaries. If haematuria is pronounced, urine can be reddish-brown (the colour of meat wastes). Microscopy of the urinary sediment susually reveals the presence of casts (mainly hyaline casts) and and cells or renal epithelium. The nitrogen excretory function of the kidneys is usually not affected in acute nephritis. Nitrogenous slags can only accumulate in the blood in serious cases attended by anurea. The clearance tests reveal more or less considerable reduction of glomerular filtration.

The infectious allergic character of acute glomerulonephritis is confirmed by immunological shifts: the content of a_2- and g-globulins in the blood incerases during the actue period.

Acute glomerulonephritis often proceeds without pronounced symptoms which make it difficult to identify it and hence to prescribe the appropritae treatment. But mild and indistinct forms of glomerulonephritis, like acute forms of this disease with classical clinical symptoms, give rise to chronic glomerulonephritis, unless the apprpritate therapy is given.

The gravest and even dramatic (Tareev) complication of acute glomerulonephritis is renal eclampsia which occurs in 4–10 per cent of patients (mostly in children and women). During a convulsive attack, the patient may be heavily contused or his ribs may be fractured. Cases were reorted where patients died from cerebral circulatory disorders or lung oedema; true, such cases are rare. Attacks of eclampsia usually leave no consequence. It is interesting tonote that eclampsia sometimes serves as a stimulus to a rapid regress of the disease and patients's recovery.

Course

Acute glomerulonephritis usually lasts only a few weeks or months. The first sing of beginnin grecovery is resolution of oedema and furtehr decrease in arterial pressure. Small haematuria and proteinuria can persist for months folowing disappearance of the main symptoms. some patients do not recover completely and the disease become chronic.

Treatment

Patients with acute nephritis should be taken to hospital. It is important that the air in the ward should be warm and dry; drafts should be absent. Sodium chloride intake should be restricted to 0.501.5 g a day, which promotes resolution of oedema and normalization of arterial pressure. Protein intake should be slightly decreased (at the expense of meat protein).

Prednisolone and other corticosteriod hormones having anti-allergic and anti-inflammatory properties are efficacious means of pathogenetic threapy of acute nephritis, Rauvolfia is given to control hypertensiopn; furocemid and otherr diuretics should be given to remove oedema.

Prophylaxis

Hardening of the body and also thorough sanation of the infection foci (carious teeth, chronic tonsilitis, sinusitis, and the like) are required.

Chronic Glomerulonephritis

Aetiology and pathogenesis

Chronic diffuse glomerulonephritis is a relatively common disease. It is often secondary to the acute form of this disease if thepatient is not timely and properly treated. In other patients, chronic glomerulonephritis occurs suddenly, without acute nephritis in their anamnesis, but it can suspected that the chronic disease was preceded by acute nephritis which however was latent, without manifest symptoms, and therefore not identified in proper time.

Chronic diffuse glomerulonephritisi can sometimes be secondary to nephropathy of pregnancy which was not treated properly. Chronic nephritis is one of the three classical forms of Britht's disease.

Great importance is now attached to the auto-immune mechanism in the pathogenesis of chronic glomerulonephritis. Antibodies to altered proteins of the renal tissue are probably formed in patients with teh disease, in addition to formation of antibodies to streptococcus. This maintains the inflammatory process in the kidneys and is the cause of chronic progressive course of the disease.

Pathological anatomy

The kidneys are not enlarged, or enlarged only slightly during the first period of the disease, which lasts several years. In the final stage of the disease, the kidneys are markedly diminished in size, their surface is granular, the renal tissue firim (arteriosclerotic kidney). Microscopy in chronic nephritis reveals mostly intracapillary inflammation in the glomeruli with gradual obliteration of the capillary loops an dthe capsule cavity and conversion of teh glomerulus into a scar or a hyaline node. Dystrophic changes occur in the epithelium of the renal tubules.

Clinical picture

The periods can be easily distinguished in thecourse of the disease: the fist period, when the nitrogen secretory function of teh kidneys is impaired only insignificantly (*the stage of renal compensation)*, and the second period, during which this function is affected substantially (*the stage of renal decompensation*).

The symptoms of the fist period ar the same as in acute nephritis. The patient may complain of weakness, more or less persistent headache, vertigo, and oedema. But the gravity of these symptoms is usually less significant than in acute nephritis. The disease is often asymptomatic and is only revealed accidentally, during out-patient examination. Objective studies help establish increased arterial pressure and hypertrophy of the left ventricle. Urinalysis reveals proteinuria and cylindruria. The presence of waxy casts is especially important diagnosistically. The urinary sediment usually contsins a small quantity (less frequently, considerable quantity) of leached erythrocyts. Teh blood serum cholesterol content is·increased. More or less significant hypopr-oteinaemia is observed due to permanane proteinuria.

Symptoms of the second, or final, period of the disease develop gradually due to the progressive nephrosclerosis. Low indices of clearance tests (especially insulin and PAH clearance tests) indicate decreased quantity of functioning kidney tissue. The filtration capacity of the kidneys remains unchanged for a long time and only decreases during exacerbation of the process. The concentration capacity of the kidneys gradually decreases along with the decrease in teh specific gravity of the urine. Removal of nitrogenous slags from the body is maintained during this period by polyuria (evacuation of much liquid from the body). Nocturnal diuresis increases by teh compensatory mechanism as well: it is two thirds-one half of the daily diuresis (nycturia). As the concentration capacity of the kidneys is affected to a greater extent, the specific gravity of the urine becomes low and its variations between 1.009 and 1.011 during the course of the day (an dunder the effect of dry food) are insignificant (isohyposthenuria). The content of nitrogenous slags in the blood of patitns (urea, creatinine, indican) increases during this period.

Symptoms of uraemia develop: weakness becomes more considerable, the patient complains of lassitude, headache, nausea, skin itching, unpleasant ammonium breath, and impaired vision. Not long before death, the patients develop uraemic coma.

Course

Chronic nephritis usually lasts from 2–3 to 10–15 years. The first period of the disease (renal compensation) is long; the second period (decompensation) is shorter. During the course of the diseae, there occur more or less prolonged periods of exacerbation, which are usually provoked by cooling or infections; exacerbations are followed by remissions.

Several clinical forms of chronic glomerulonephritis are differentiated by the character of its course adn prevailing symptoms. The *nephrotic form* is characterized by oedema, the urinary syndrome,and a comparatively rapid course. The hypertensive form is comparatively benign and is characterized by teh hypertensive syndrome and insignificant chanbes in the urine. The mixed form is characterized by oedema, changes in the urine, and arterial hypertension. This form of glomerulonephritis is the gravest and comparatively rapid: a pronounced renal failure develops in 2–3 years. Finally, there is the latent form of the disease, which is not manifested by oedema or pronounced hypertension; the changes in teh urine are only insignificant; renal failure develops at late terms, often only in 10–15 years. As a rule, the patient dies of renal failure.

Treatment

Patients with exacerbaions are prescribed bed-rest, a dairy and vegetable diet with restricted sodium chloride (to 1.5–2.5 g/day) and containing at least 1 g/kg protein. The daily protein intake in the nephrotic form and hypoproteinaemia should be slightly increased. Foci of chronic infection (carious teeth, tonsillitis, etc.) should be treated. Infectious foci are treated with antiboiotics. Good effect in the treatment of exacerbated nephritis (nephritic form of diffuse glomerulonephritis) is attained with prolonged use of chloroquine diphosphate (in the absence of marked hypertension or azotaemia). The course of treatment continues for several months. Stable clinical remission and even recovery of patients can sometimes be obtained with this therapy.

Symptomatic therapy in hypertensive and oedematous forms of chronic nephritis includes hypotonics (reserpin) and diuretics (hypothaiazide, furocemid). Sanatorium therapy is often very helpful to patients with hypertensive and nephrotic forms of glomerulonephritis with compensated renal function.

Control of azotaemia is important in the treatment of uraemia. The intake of animal protein (meat) should be limited to 18–30 g/day. Broad spectrum antibiotics and also sour milk products (yoghourt, sour milk) are given to inhibit the putrefactie process in the intestine. In the absence of tendency to oedema, ample liquid is indicated. Group B vitamins, ascorbic acid, glucose, repeated blood transfusions, gastric lavage with sodium hydrocarbonate, sodium hydrocarbonate enema are used to control toxicosis. Peritoneal dialysis and haemoldialysis (artificial kidney) are more effective means to control uraemic toxicosis. These means do not remove the cause of uraemia but only prolong the patient's life. More prospective treatment of severe forms of chronic nephritis is translplantation of the kidneys.

Prophylaxis

Prophylaxis consists in timely treatment of acute and chronci focal infections (tonsillitis, sinusitis, carious teeth, paradontosis, etc.). Patietns with chronic glomerulonephritis should be regularly inspectred.

Toxic Kidney

Toxic kidney (acute nephrotic syndrome, acute nephrosis, necronephrosis) occurs in acute infectious and toxic diseases, such as typhus, malaria, influenze, ingestion of some nephrotoxic substances (corrosive sublimate, carbon tetrachloride), in transfusion of incompatible blood, massive burns, and some other cases.

Pathological anatomy

In milder cases, insignificant dystrophy of epithelim of the proximal tubules occurs in the form of opaque swelling and fat infiltration in teh tubular cells. In sever cases, thekidneys are slightly enlrged and flaccid. Inpoisoning with corrosive sublimate, the kidney is first red (large red kidney) due to pronounced plethora. Its vessels are then affected by the spasm and the kidney contracts (small white kidney). Histology reveals proteinous dystrophy and necrosis of the tubular cells and thier desquamation; the glomeruli are often affectd as well (acute necronephrosis).

Clinical picture

Symptoms of the disease vary greatly. Mild forms are practically asymptomatic; protein can only be detected in the urine. Febrile albuminuria occurring in infectious diseases can be used as an illustration. In severe forms of the disease, the urien excretion is upset, and oliguria develops. Hypertension adn oedema are absent in typical cases. The urine is concentrated and its specific gravity is high; it contains protein, various casts, erythrocytes, cells of renal epithelium,and leucocytes. Renal dysfunction is aggravated by the disordered haemodynamics which attends shock (e.g. in burns, injuries). Anuria attends most severe cases; nitrogenous slags are accumulated in the blood. The patient dies within a few days unless excretion of ureine is not restored.

Treatment

The main disease whichis aggravated by the kidney affection should be treated. If olgurea or anurea persits, peritoneal dialysis or haemodialysis are indiacated to prevent uraemia.

Amyloidosis of the Kidney (Amyloid Kidney)

Renal amyloidosis (amyloid nephrosis, amyloid dystrophy of teh kidneys) is manifestationof the general disease, amyloidosis.

Aetiology and pathogenesis

Congenital (herditary), primay and secondary congenital amyloidosis are distinguished. Secondary amyloidosis develops in the presence of of pronounced protein metabolic disorders, mostly in patients with chronic inflammatory diseases such as tuberculosis, osteomyelitis, or bronchiectasis. Less frequently amyloidosis develops in deforming polyarthritis, lymphogranulomatosis, and some other diseases. It is believed that in all these diseases, in connection with chronic action of infection and toxins, and also decomposition of tissues and leucocytes, synthesis of prteins is distorted in the reticuloendothelial system. The auto-immune mechanism is acutated at a certain stage, and the production of antibodies to the altered own proteins is intiated. The antigen-antibody complexes, in the form of a specific firm amyloid substance, which is a combination of globulin and mucopolysaccharides, are deposited in various organs and first of all in the walls of fine vessels and capillaries under the argyrophilic membrane, under the tunica propria of teh glands, and in the reticular stroma of various organs. Metabolism in the adjacent cells is upset due to deposition of amyloid. The cells and other tissue elements become compressed and atrphied. Genetically determined amyloidosis is the result of congenital defects of the enzyme systems responsible for theprotein synthesis. Aetiology if sporadic primary amyloidosis is uncertain.

Pathological anatony

Insignificant changes are found in the spleen, kidneys, liver adrenal glands, and teh gastro-intestinal tract (less frequently in lymph nodes and other organs) in amyloidosis. Amyloid kidneys are enlarged, consolidated, and grey ("large fatty kidney"). Histology reveals amyloid grains under the argyrophilic membrane of the arteries andcapillaries. Capillary loops become obliterated, the glomeruli are replaced by amyloid grains or connective-tissue scars. Dystrophic changes in the tubular epithelium are also observed. Amyloid-affected (contracted) kidney is the result of a severe process.

Clinical picture

The clinical picture of all types of amyloidosis is the same. The clinical piture of secondary amyloidosis is largely determined by the main disease (pulmonary tuberculosis, osteomyelitis, etc.) and also (in all types of amyloidosis) by the degree of affection with amyloidosis of the other organs, e.g. the gastro-intestinal tract, the liver, the heart,and other organs and systems. Amyloid kidney usually occurs with the nephrotic syndrome, which has its effect on the clinical picture of the disease. In the initial period of the disease the patient's general condition is not affected. The study of the urine can only reveal proteinuria (to 10-15 g/l). As distinct froma "purely" nephrotic syndrome, the urine contains not only albumins but also globulins. Single casts (hyaline, granular, waxy, epithelial cells, leucocytes) can be found in the urinary sediment. The erythrocyte content is usually low, but some patietns have a pronounced nephrotic syndrome with typical changes in the urine.

The blood study reveals hypoproteinaemia with a specific prevalence of globulins (the albumin to globulin ration decreaeses to 1.0 and more), and with the prevanelce of a2- and g-globulins. Hypercholesterolaemia may also occur.

The differential diagnosis between amyloid kidney and othe rchronic diseases of the kidneys is often facilitated by teh test with intravenous administration of Congo red. The method with subcutaneous injection of 1 ml of a 1 per cent methyl blue solution is commonly used. One-hou specimens of the urine are taken during 5 to 6 hours following administration. Normally, all specimens should be coloured green. In the presence of amyloidosis, the colour changes insignificantly or does not change at all. But negative tests with Congo red or methylene blue do not rule out the presence of amyloidosis because the degree of affection may be insufficient to ensure reliable retention of the dye. An accurate diagnosis of amyloidosis can only be established by the results of nephrobiopsy.

Persistent oedema develops at later periods of the disease. Despite teh affection of the renal vessels arterial hypertension is a rare symptom in amyloidosis. Glomerular firltration in patitnts with amyloid kidney becomes upset during tehearly period. At the final stage, a picture of pronounced renal failure arises; it is caracterized by azotaemia and uraemia, from which the patient usually dies. Less frequently the patient dies of cachexia or other causes.

Treatment

Treatment of secondary amyloidosis can only be successful if teh main cause of the disease is removed before severe changes have taken place in the organs, in the kidney in the first instance. Treatment includes a protei-rich diet (provided the nitrogen-excretory function of the kidneys is satisfactory) and containing limited quantity of sodium chloride; the therapy also inclusdes vitamins and control of oedema and azotaemia (in the presence of renal failure).

Prophylaxis

It consists in timely revealing and treating chronic inflammatory processes and purulenet diseases to which amyloidosis is secondary.

Nephrolithiasis

Calculi are formed in the renal pelves in nephrolithiasis. The chemical composition of calculi is quire varied. They usually contain phosphates (calcium and magnesium salts of phosphoric acid). Less frequently calculi consisting of salts of oxalic acid (oxalates), uric acid (urates), and carbonic acid (carbonates) are encountered. Calculi may contain proteins, xanthine, cystine, and sulphonamide.

Aetiology and pathogenesis are uncertain. But it has been established that formation of calculi is stimulated by the infection of the urinary tract, injuries to the kidneys, and haemorrhages into the renal tissue, urinary congestion, and some avitaminoses (A, D). Metabolic disorders (hyper-parathyrodism) and sharp changes in the pH of the urine also promote calculi formation.

It is believed that precipitation of salts from the urine and calculi formation occur around an organic "nucleus" which may be desquamated cells of pelvic epithelium, accumulation of leucocytes, a blood clot, etc. But salts may precipitate if their concentration in the urine is increased or their solubility decreases due to chanages in the pH of the urine, or if the concentration of the socalled protective colloids in the urine, which ensure stability of supersaturated solutions, decreases. For example, concentration of the uric acid in the urine is usually 15–20 times higher than its solubility in water.

Clinical picture

The disease is characterized by attacks of nephrolithiasis (renal colic) whcih are followed by interparoxysmal periods.

Most patients do not complain of anything in the interparoxysmal period. Only dull pain in the lumbar region occurs in some of them. The Pasternatasky symptom is as a rule positive. The study of the urine sometimes reveals transient haematuria; salt crystals are often found.

The first sign of the disease is usually an attack of renal colic, which develops during thepassage of a calculus via the ureter. The attack occurs suddenly, often after jolting or long walking. The pain is localized in the lumbar region with radiation downward, by the course of the ureter, and into the sex organs. The pain is very severe and the patient is restless; he changes his posture continually. Pain intensity may lessen for a short time but then it intensified even to a greater extent. The attack is attended by frequent and painful urination and various reflex symptoms (nausea, flatulence of the abdomen, retained stools). Erythrocytes and protein are found in the urine. The attack ends when the calculus passes into the urinary bladder. Sometimes the calculus passes the urethra and is excreted. Attacks recur at various intervals, from several attacks during one month to one during many years.

In the presence of typical attacks of *renal colic*, it is easy to diagnose *nephrolithiasis*. But it is sometimes difficult to differentiate renal colic in the right side from an attack of pain occurring in cholelithiasis. It should be remembered that pain in attacks of renal colic radiates downwards, whereas in attacks of hepatic colic it radiates upwards, into the right shoulder, sholder blade, and the diuretic symptoms ar absent. An attack of hepatic colic may end with jaundice.

Diagnosis in remission can be facilitated by X-ray examination (*pyelography*). The claculus can rarely be revealed on a survey X-ray picture of the kidneys. Oxalats, phosphates, and carbonates give an intense shadow on the X-ray patterns.

Course and complications

Long-stnding presence of concerements in the renal pelvis has its specific effects: *pyelitis* (inflammation of the renal pelvis) usually develops, which can later transform into pyelonephritis.

If a calculus is retained in the ureter to obliterate its lumen, the renal pelvis becomes distended by the accumulated urine to cause hydrops of the kidney (*hydronephrosis*) which causes future atrophy of the renal tissue. *Pyonephrosis* (acute purulent inflammation of the pelvis with involvement of the kidney tissue) develops if urine is infected. Pyonephrosis is characterzed by a grave general condition of the patient, hectic fever with profuse sweating, dull pain in the corresponding side of the loin, neutrophilic leucocytosis with shift to the left,and markedly increased ESR.

Treatment

All measures should be taken to remove spasm and pain in attacks of renal colic. A hotwater bottle should be placed on the loin and atropine given subcutaneously. Hot water baths are also helpful. Procaine block in the lumbar region is given.

In the interparoxysmal period patients are recommended to drink much liquid. In the presence of urates, alkaline mineral water is desirable (Borzhomi, Essentuki mineral water). Great importance is given to the diet. Food rich in calcium chloride (milk, curd cheese, potatoes) and other substances that compose stones should be taken in limited quantity. In the presence of oxalates, green lattuce, beans, tomato and other foods containing oxalic acid should be excluded from the diet. If urtes are present, meat, fish and foods rich in purine bases should be ruled out.

If small stones are found in the pelvis, long walks, ample liquid intake, and antispasmodics containing essential oils (enathine, cystenal) should be prescribed to stimulate the excretion of the stones. If pylenophritis joins the process, antibiotics and nitrofuranes should be given. Infected large stones should be removed surgically. This, however, does not eliminate the cause of stone formation and they can be formed again.

PYELONEPHRITIS

Acute Pyelonephritis

Aetiology and pathogenesis

Acute pyelonephritis arises as a result of infection spreading from the renal pelvis onto the kidney tissue in acture pyelitis, or as a result of infection of the kidney or its pelvis via haematogenic ruote in the presence of infectious foci in the patient's body (*chronic tonsillitis, osteomyelitis*). It may also develop during acute infectious diseases (acute tonsillitis, sepsis, typhoid fever, etc.). Penetration of the bacteria into the kidney and the pelvis does not always provoke acute pyelonephritis or focal nephritis: bacteriuria without symptomsof involvement of the renal tissue is often observed. Difficult urine outflow from the kidney (stone in the ureter, twisting of the ureter) stimulates development of *pyelonephritis*.

Pathological anatomy

The kidney is slightly enlarged, mucosa of the renal pelvis is inflammed and oedematous, ulceration is possible. If urine out flow is obstructed, the pelvis contains pus. Inflammatory infiltration of the renal tissue develops and purulent foci can be found in some parts of the kidney.

Clinical picture

The patient's condition is grave: high irregular fever, chills, dull lumbar pain. If inflammation spreads over onto the urinary bladder and the urethra, the patient feels tenesmus and sharp pain during urination.

Study of the urine reveals *pyuria* and *bacteriuria*. In complete obstruction of the ureter and unilateral affection there may be no changes in the urine. Obliterated forms of acute *pyelonephritis* may also be observed (usually in the presence of grave general diseases). In such cases the kidney affection can only be suspected form the urinalysis.

Treatment

Antibiotics, sulpha and nitrofuran drugs (furadonin) are recommended.

Chronic Pyelonephritis

Aetiology and pathogenesis

Pyelonephritis often arises in patients with chronic pyelitis due to transition of the inflammatory process from the renal pelvis onto the renal tissue. *Nephrolithiasis* facilitates fixation of the infection in teh pelves and its spreading onto the renal tissue. chronic pyelonephritis is usually caused by conventionally pathogenic flora, intestinal escherichia; less frequently the disease is provoked by enterococus, Proteus, or other infection. As infection spreads from the renal pelvis onto the renal parenchyma, the papillae are first affected, and then the medullar and cortical layers of the kidney. The kidney contracts as a result. The process is often unilateral. If both kidneys are involved, the extent of affection may differ.

Clinical picture

Involvement of the renal tissue in the process alters the clinical picture of *pyelitis* to make it similar to that of *chronic glomerulonephritis*: hypertension develops, the concentrating and nitrogen excreting functions of the kidneys are gradually deranged, and uraemia develops.

But as distinct form chronic glomerulonephritis, pyelonephritis is characterized by involvement of only one kidney or asymmetrical affection of the kidneys. This can be revealed by intravenous or retrograde pyelography, separate study of the urine from the right and left kidneys obtained by catheterization of the ureters, and also by separate studies of some substances excreted by the kidneys. Clearnce tests are useful, since they can reveal early signs of involvement of the distal tubules by the delayed excretion of phenol red and decreased excretion coefficient. Radiographic methods, such as renography and scanning, are also used for the purpose.

Pyelonephritis is characterized by the signs of infectious inflammation of renal pelves, i.e. by the presence of *bacteriuria, leucocyturia* (especially the presence of active leucocytes known as Sternheimer-Malbin cells in the urinay sediment), and also deformation of the renal pelves as revealed by pyelography. Bacteriological study of the urine is of great significance: (cultivation of the urine on a nutrient media, and determination of bacterial sensitivity to antibiotics). It should be remembered that urine specimens from women should onlybe taken by catheterization of the bladder. Differential diagnosis is facilitated by puncture biopsy of teh kidneys.

At the terminal stage, due to involvement of both kidneys and development of nephrosclerosis, all functional tests for kidneys are not informative. Death of patients is in most cases caused by uraemia.

Treatment

Infection is treatedby antibiotics, sulpha drugs, and nitrofuran derivatives. Symptomatic therapy is given to control *hypertension* and *azotaemia* (at the terminal stage).

5

DISEASES OF THE BLOOD

SPECIAL PATHOLOGY

Anaemia

Anaemia is a pathlogical condition characterized by decreased number of erythrocytes and/or haemoglobin content in a blood unit volume due to their general deficiency (Gk an not, haemia blood,i.e. deficient of blood).

Anaemia should be differentiated from hydraemia (abnormally watery blood) in which the erythrocyte and haemoglobin are deficient as well, but not at the expense of their absolute reduction but due to the dilution of blood in renal, cardiac and other oedema. Anamia should also be differentiated from oligohaemia, which is the reduction of the total volume of blood, e.g. immediately after a profuse haemorrhages. The total mass of circulating blood can be normal in anaemia (normovolaemia), increased (hypervolaemia) orr decreased (oligohaemia or hypovolaemia). Thickening of blood in persistent vomiting and profuse diarrhoea can mask anaemia because the total amount of plasma decreases and the number of erythrocytes and haemoglobin in a unit volume of the circulating blood can be normal or even increased.

Anaemia is often characterized not only by quantitative changes in the red blood composition, but also qualitative changes in the structure of erythrocytes and haemoglobin molecules. These changes are important for the transport function of blood and tissue respiration, and can be the cause of additional pathological changes in the body. For example, a congenital defect of erythrocytes in some hereditary haemolytic anaemia may (Due to their intense haemolysis) cause haemosiderosis of the internal organs, formation of pigment stones in the gall bladder, etc.

Anaemia has a pronounced effect on the vital activity of the boyd. Anaemization causes oxygen hunger of organ and tissues (hypoxia) and their dystrophy. For example, if the blood haemoglobin content is halved (70-80 g/l),1 initial symptoms of myocardial dystrophy develop. If the haemoglobin content decreases to 50 g/l, the dystrophic changes become prounounced. Unoxidized products of metabolism (lactic acid, in the first instance) accumulate in the body due to hypoxia. The alkaline reserve of blood decreaes. In grave cases, a tendency to acidosis develops which causes further dystrophy of tissues. Severe anaemias attended by marked disorders in tissue metabolism are incompatible with life.

Anaemia of any origin is accompanied by some compensatory processes, which partly remove or lessen its consequences: (1) blood circulation is intensified, i.e. stroke and minute volumes increase, tachycardia develops, and the rate of blood flow increases; (2) blood distribution is altered, blood depots in the liver, spleen, and muscles are activated, and the blood supply to theperipheral tissues becomes limited at the expense of the increased blood supply to the vital organs; (3) oxygen utilization in tissues is intensified and the role of anaerobic processes in tissue respiration increases (anaerobic respiration with glutathione); (4) the erythropoietic function of bone marrow is stimulated. More than 50 types of anaemia are now differentiated.

According to their origin, the following types of anaemia are distinguished.

1. *Anaemia* due to loss of blood (Acute and chronic).

2. *Anaemia* due to disordered (haemopoiesis) in deficiency of iron (necessary for the production of haemoglobin), in vitamin B12 deficiency (necessary for normal erythropoiesis), in inhibition of the bone marrow by endogenous or exogenous toxicosis, radiation, or by some unknown factors, and also in cases where red bone marrow is replaced by other tissues, e.g. myeloma or multiple metastases.

3. *Anaemia due to excessive haemolygis.* This type of anaemia is sub-divided into: (a) anaemia with prevalent extravascular (intracorpuscular) haemolysis of erythrocytes in macrophages of the spleen, and,

to a lesser extent, in the bone marrow and liver. These are anaemia caused by herediatary morphological and functional erythrocyte deficiency (*spherocytic* and *ovalocytic anaemia*), and *auto-immune haemolytic anaemia.* They are all characterized by hyperbilirubinaemia and splenomegaly;; (b) Anaemia with intravascular, usually acute haemolysis (in anaemia) attended by release into the plasma of unbound haemoglobin and by haemoglobinuria; haemosiderosis of the internal organs is observed also in chronic haemolysis (e.g. in *Marchiafava-Micheli disease*). This classification is only conventional because both intracorpuscular and vascular haemolysis can occur in one and the same form of haemolytic anaemia.

Haemolyic anaemia is also often subdivided as follows: (a) *hereditary* (*Congenital*) *anaemia*, which includes membranopathy of erythrocytes (Associaed with abnormality of protein or lipid complex of erythrocyte envelope, causing changes in their shape and premature decomposition; microspherocytic anaemia, ovalocytic anaemia, etc.); enzymopenic which promotes their accelerated decomposition) and haemoglobinopathy in which the structure of haemoglobin or its synthesis are disturbed (sickle-cell anaemia, *thalassaemia*); (b) *acquired anaemia* (*auto-immune haemolytic*) and *iso-immune anaemia*, and also anaemia caused by mechanical injury to erythrocytes, acquired membranopathies, toxic anaemia, etc.).

Apart from the pathogenetic classification, there are classifications based on other principles. Three groups of anaemia, for example, are distinghished in accordance with haemoglobin saturation of erythrocytes (by the colour index); *normochromic* (0.8-1.0), *hypochromic* (less than 0.8) and *hyperchrromic* anaemia (more than 1.0). The group of hypochromic anaemia includes iron-deficiency anaemia: chronic (less acute) posthaemorrhagic anaemia, gastrogenic iron-deficiency anaemia, and juvenile chlorosis. Hyperchromic anaemia is caused by the deficiency of vitamin B_{12}. This is *Addison-Biermer anaemia, bothriocephalus anaemia,* and also *achrestic anaemia* (Due to defective utilization of vitamin B_{12}). Other anaemias proceed without considerable changes in the colour index of blodo and are therfor normochromic.

It is very important to assess the regenerative capacity of the bone marrow upon which (to a certain degree depend treatment and prognosis of the diseaes. Distinguished are regenerative anaemia, i.e. anaemia in which the bone marrow preserves its capacity to produce new erythrocytes; hyporegenerative anaemia, in which this capacity is impaired; and aregenerative or aplastic anaemia, in which bone marrow function is completely or almost completely lost. The regenerative function of the bone marrow is assessed by the rate at which th equantity of reticulocytes in creases in the peripheral blood and by the proportionof the erythro-and leucoblastic elements in the sternal punctate. Their normal ratio is 1 : 3 or 1 : 4, while in regenerative anaemia, in which erythropoiesis dominates in the compensatory function of the bone marrow, this ratio becomes 1 : 1, 2 : 1 and even higher. This shift is absent in hydro- or aregenerative anaemia while the reticulocyte content of the peripheral blood is low.

Acute Posthaemorrhagic Anaemia

Anaemia caused by an acute blood loss (*Acute posthaemorrhagic anaemia*) occurs mostly in various injuries associated with traumatized large vessels (extrauterine pregnancy, delayed placental detachment during labour, etc.). Acute posthaemorrhagic anaemia occurs in diseases that can be attended by profuse bleeding e.g. in gastric an dduodenal ulcer, degrading tumour of the stomach, kidneys, or thelung, in tuberculosis and abscess of the lung, bronchiectasis, varicose dilation of the oesophageal veins in liver cirrhosis, haemorrhagic diathesis, and especially in haemophilia.

Clinical picture

In cases with external haemorrhage, the physician can often locate the source of bleeding at first sight (E.g. in injury). The patient' grave condition can in these cases be directly attributed to profuse blood loss. Haemorrhage from the internal organs can be manifested by blood vomiting (unaltered blood originates) from the oesophagus; brown blood from the stomach), by expectoration of blood (Scarlet foaming liquid), by the presence of blood in faeces (melaena in haemorrhage from the stomach or the small intestine; dark or scarlet blood originates from the large intestine, especially from its terminal part) and by blood presence in the urine (haematuria). It should be remembered that in gastro-intestinal haemorrhage, the blood can only be discharged into the enviornment in a certain lapse of time (with the vomit or excretions). Moreover, haemorrhage caused by the rupture of the spleen, liver, or by the internal injury to the chest can be difficult to establish because blood will accumulate in the abdominal or pleural cavity.

The first sign of a sudden haemorrhage is the feeling of weakness, dizziness, noise in the ears, palpitation of the heart, nausea, and in rare cases vomiturition. In severe cases withprofuse blood loss, the patient is in the state of shock (if the bleeding is caused by an injury) or collapse (if haemorrhage is due to infection of the internal organs). The patient's condition depends not only on the amount of blood loss, but also on the rate at which blood is lost. Inspection reveals pronounced and in some cases deadly pallidness; the skin is covered with sticky cold sweat, the skin temperature is subnormal. Respiration is superficial and accelerated. The pulse is fast, small, and (in severe cases) thready. Arterial pressure (both systolic and diastolic)) is low. Ausculation of the heart reveals marked tachycardia.

The pathological and compensatory changes in acute blood loss with benign outcome can be divided into three stages (or phases). First oligohaemia develops. It causes a reflex spasm of the vessels to decrease the volume of the vascular system and to recover blood from its reserves (depots). For this reason, the blood haemoglobin and erythrocyte content may remain normal within the first hours (or even within 1 or 1.5 days) following the blood loss. Tissue fluids are drawn into the vessels to cause hydraemia in 2 or 3 days: the erythrocyte and haemoglobin content in unit voluem decreases. Signs of marked activation of erythropoiesis appear on the third to seventh day. Anaemia becomes hypochromic in the loss of considerable amount of blood due to exhaustion of the iron store.

Treatment

Bleeding should be arrested as soon as possible by placing a tourniquet or by a tamponade of the external haemorrhages in wounds. Surgical intervention is indicated in continuing haemorrhage from the internal organs. Measures to prevent shock or collapse should also be taken. Blood loss should be compensated for by infusion of whole blood or its substitutes; cardiac and vascular medicinal preparations should be administered. Iron preparations should be given to patients with profuse blood loss in several days after the haemorrhage has been arrested.

Iron Deficiency Anaemia

Iron deficiency anaemia (anaemia sideropriva gastroenterogenica) arises in the deficit of iron whichis necessary for the productionof haemoglobin in erythrocytes. This type of anaemia develops in patients with decreased iron absorption due to resection of the stomach ("agastric anaemia"), removal of a considerable part of the small intestine, especially of its proximal part, in intestinal diseases attended by abnormal absorption, and in the iron deficit in food. The latter occurs mosty in children with prolonged milk diet and copper deficit. Increased iron demands occur during intense growth of the body. During establishment of the menstrual cycle in young girsl (menstrual loss of blood and iron ions) juvenile iron deficiency anaemia (juvenile chlorosis) may develop. Chronic haemorrhage also causes iron deficiency anaemia.

Repeated (not profuse) loss of blood cause anaemization due to exhaustionof theiron store which is necessary for the productionof haemoglobin in erythrocytes. Daily intake of iron with food is small, about 11–28 mg, and one fourth of this quantity is only absorbed. This is equivalent to theiron content of 15 ml of blood. Daily loss of 15 ml (or even smaller amount) of blood therefore inevitably exhausts the iron store to cause iron deficiency anaemia.

Chronic blood loss and chronic posthaemorrhagic anaemia attends many diseaes of the internal organs, and in the first instance, of the gastro-intstinal tract. In most caes these are gastric or duodenal ulcer, cancer, polyposis of the stomach and intestine, haemorhoids, and certain types of helminthiasis. Chronic posthaemorrhagic anaemia often occurs in tumour of the kidneys, cavernous tuberculosis of the lungs, and in uterine haemorrhage.

Some other factors promote anaemia. These are mainly those factors which can decrease the iron stores of the body. For example, patients with secondary gastric hyposecretion and enteritis develop anaemia sooner and it runs a more severe course in the presence of even insiginficant chronic haemorrhage. Gravity of chronic posthaemorrhagic anaemia arising in patients with degrading tumours of the gastro-intestinal tract, kidney, or the uterus, is intensified by the toxic effect of the tumour on the haemopoiesis and by multiple metastases into the bone marrow, etc. Hydrochloric acid of the gastric juice promotes reduction of trivalent iron to its divalent form, which is easier assimilated. But recent studies show that hydrochlorric acid does not play a decisive role in activation of iron abosrption.

In the absence of adequate iron supply to the body or its utilization from the store, the synthesis of haemoglobin, myoglobin, and iron-containing enzymes of various cells involved intehoxidation processes is upset. This impairs nutrition of tissues and accounts for the development of many symptoms of the disease. The clinical picture or iron deficiency anaemia is explained by insufficient oxygen transport to tissues due to anaemia on the one hand, and by disordered cell respiration on the other.

Clinical picture

Slow development (within months and years) of iron deficiency anaemia accounts for actuation of the compensatory mechanisms. Most patients therefore are well adapted to the disease and can satisfactorily stand even significant anaemia.

We shall not discuss patient's complaints associated with the main diseaese, to which anaemia is secondary (e.g. the cause of chronic haemorrhage). The specific complaints of anaemia patients will only be emphasized: weakness, dizziness, dyspnoea (especially exertional), increased fagigue, nosie in the ears, and fainting. Many patients develop various dyspeptic symptoms: decreased appetite, preverted taste, slight nausea, heaviness in the epigastrium after meals, and regurgitation. Diarrhoea is also frequent. Slight paraesthesia (tingling and pricking) is possible. Excrciating dysphagia sometimes develops durign swallowing dry or solid food in especially severe cases. This sideropenic dysphagia was first described by Rossolimo and Bekhterev in 1900–1901. Later this syndrome was described by Plummer and Vinson. The dysphagia is explained by extension of the atrophic process from the stomach onto the oesophageal mucosa, and osmetimes by tis development in the proximal part of the soft connective-tissue membranes and bridges.

Inspection of the patient reveals pallor. Certain trophic changes in the skin, its appendags, and mucosa can be due to the general iron deficit. The skin is dry and sometimes slightly scaling. The hair is brittle, early grey, and showing the tendency to falling. The nails become flat, sometimes spoon-like, opaque, marked by transverse folds, and brittle (koilonychia). The mouth angles often have fissurs (angular stomatitis), the papillae of the tongue are levelled (atrophic glossitis). The teeth lose their luster and quickly decompose despite a thorough care. If iron preparations are taken for a long time, the teeth may blacken ue to formation of black iron sulphite (by the reaction of iron with hydrogen sulphide which is liberated by the carious teeth). Purulent inflammation of the gum mucosa around the tooth necks develops (alveolar pyorrhoea).

Physical examination can reveal a slight indistinct enlargement of the left ventricle, sytolic murmur at the heart apex, an nun's murmur over the jugular vein (mostly on the right). Lymph nodes, liver and spleen are not enlarged.

Study of the blood reveals decreased erythrocyte and even more decreased haemoglobin content of the blood. The colour index is less than 0.85; in grave caes it is 0.6–0.5, and even lower. Microscopyof blood (Plate 30) reveals pallid erythrocytes (hypochromia), anisocytosis, and poikilocytosis. The average diameter of erythrocytes is less than normal (microcytosis). The number of reticulocytes is small. Anaemia is usually attended by thrombocytoleukopenia, sometimes relative monocytosis, lymphocytosis, and eosinopenia. The iron content of the serum is decreased (1.5–2.5 times and more). The percentagae of transferrin saturation also decreases.

Decreased activity of the iron-containing enzymes of tissue respiration provokes (or intensifies) atrophy of the gastro-intestinal mucosa. The study of gastric juice reveals in most cases aachlorhydria or even achylia; the total amount of the excreted juice is much decreased. X-rays reveal levelled folds of the oesophageal and gastric mucosa. Oesophagoscopy and gastroscopy confirm atrophy of the oesophageal and gastric mucosa.

Course

The course of the disease is chornic and gradually progressive if the iron deficiti in the body increases.

Treatment

Iron preparation (haemostimulin, etc.) are given. If the patient has gastritis or peptic ulcer, the iron preparations should better b givne intramuscularly or intravenously (ferbitol, fercoven, etc.). The therapy gives a comparatively rapid and permanent effect: work capacity is rapidly restored, erythrocyte and haemoglobin of blood normalize in 3–5 weeks. The patient should however be regularly (several times a year) given prophylactic courses of therapy withiron preparations in order to prevent possible relapses of the disease. The diet of patients with iron deficiency anaemia should be rich in iron salts, e.g. liver, meat, eggs, apples, dried fruits. Efficacy of treatment of anaemia caused by chronic loss of blood depends on removal of the source of blood loss.

Vitamin B_{12} (Folic Acid) Deficiency Anaemia

Aetiology and pathogenesis

Vitamin B_{12} (folic acid)) deficiency anaemia was first described by Addison in 1855. One of its forms was later given the name of Addison-Biermer anaemia. In 1868, Biermer published a more detailed description of the disease, which he called pernicious or malignant anaemia, because its prognosis was then grave and patients usually diet in a few months or years after the appearance of the fist symptoms.

The disease was effectively treated for the first time by Minot and Murphy. The patietns were given raw calf liver in large amounts every day. Minot and Murphy noted that distinct remissions followed in patients who were given this diet and conjectured that raw liver contained a certain substance which is necessary for normal *haemopoiesis*, and whose absence or deficit causes pernicious anaemia. The next stage in the study of this disease is connected with experiments carried out by Castle. He noted that meat treated with gastric juice (containing various amounts of the stomach of patietns with the*Addison-Biermer anaemia*. Gastric juice alone, untreated meat, or meat treated with gastric juice of patients with the Addison-Biermer anaemia have no anti-anaemic effect. Castle suggested that a special substance, haemopoietin, was necessary for normal maturation of erythrocytes. Haemopoietin is produced by combination of a certain extrinsic factor supplied with food and the intrinsic factor contained in normal gastric juice.

At the present time, Castle's conjecture concerning the pathogenesis of Addison-Biermer anaemia has been proved experimentally and clinically. The extrinsic and intrinsic factors and their biological role have been studied sufficiently well. The factor is vitamin B_{12} (*cyanocobalamin*) discovered by Smith in 1948 which is contained in calf liver, kidneys, meat, eggs, and gastromucoprotein produced by the accessory cells of the glands found in the fundus of the stomach. In healthy subjects, vitamin B_{12} combines with gastromucoprotein in the stomachto give a sufficiently stable complex which protects vitamin B_{12} from intestinal microflora to ensure adequate absorptionof vitamin B_{12} (mainly in the ileum). Gastromucoprotein is absent from the gastric juice of patients with the Addison-Biermer anaemia due to pronounced atrophic gastritis. In the absence of gastromucoprotein, vitamin B_{12} delivered with food is decomposed by intestinal flora an dis not assimilated by the body to cause vitamin B_{12} deflict. In other cases, vitamin B_{12} (folic acid) deficiency anaemia is the result of vast resection of the stomach, severe enteritis, increased demands for vitamin B_{12} in pregnancy, its consumption by helminths (bothriocephaliasis), and in disordered assmilation of this vitamin by the bone marow (achrestic anaemia).

An important biological effect of vitamin B_{12} is activation of folic acid. Like vitamin B_{12}, folic acid belongs to substances included in the group of vitamin B. It is contained in leaves of various plants, fresh vegetables, beans, liver, and kidneys of animals. Folic acid is deposited in the human body mainly in the liver, whre it is present in inactive stage. Vitamin B_{12} promotes formation of folic acid derivatives, folatles, which are probably the factor necessary for haemopoiesis in the bone marrow. In conditions associated with vitamin B_{12} and folate deficiency, the synthesis of DNA is disordered; this in turn causes disorders in cell division; the cells become large and qualitatively inadequate. Erythroblasts are affected most severely: large cells of embryonal haemopoiesis, megaloblasts, are found in the bonr marrow instead of erythroblasts. They are not only larger than erythroblasts; they also differ in the structure of their nuclei and protoplasm, earlier andmore intense saturation with haemoglobin during their differentiation (at the stage of reticular structure of the nucleus), related mitotic division, and mainly in their inability to grow to normal erythrocytes. Most megaloblasts are decomposed in the bone marrow before they reach the stage of a nucleated cell. Only a small quantity of megaloblasts are differentiated to anuclear cells (megalocytes) and enter the blood vessels. Megalocytes are larger and more saturated with haemoglobin than erythrocytes and differ from them by morphological and functional inadequacy. Megalocytes have no such high oxygen-transport capacity as the erythrocytes and are quickly decomposed by reticuloendothelial cells: the average life of megalocytes is about three times shorter than of erythrocytes.

The absence of gastromucoprotein in gastric juice (like achlorhydria which usually attends this disease) is due to atrophy of the gastric mucosa. Some investigators believe that atrophy of gastric mucosa is not inflammatory in its origin as it was believed earlier (*atrophic gastritis*) but is a result of congenital insufficiency of its glandular appapratus which is manifested with time. In the opinionof other authors, the atrophy of gastric mucosa is caused by antibodies produced by the patient's body to the gastric glandular cells, which can however be slightly altered bytoxic effects or inflammation (auto-imune mechanism).

If the second coenzyme of vitamin B_{12}, *deoxyadenosylcobalamin*, is deficient, fat metabolism becomes upset with accumulation of *methylmalonic acid*, which is toxic for the nervous sytem (provokes funicular myelosis).

The *Additon-Biermer anaemia* attacks commonly the aged; the incidence among women is higher than in men.

Pathological anatomy

The skin and the orgn are pallid. Small haemorrhages are possible. *Haemosiderosis* of the liver, kidneys, bone marrow and dystrophic changes in them are characteristic. These changes are observed also in the myocardium, the brain, and the spinal cord (mostly in the lateral cord). Bone marrow is affected by hyperplasia; it is bright-red, the foci of extramedullar haemopoiesis are seen in the spleen and the lymph nodes. Histological studies show prevalence of red blood cells; many young forms, myeloblasts are seen. Megaloblasts, the large cells of perverted erythropoiesis, are especially numerous.

Clinical picture

The onset of the disease is insidious. The patient grows weaker, he complains of heart palpitation, dizziness, and dyspnoea, especially during exercise or brisk movements; the work capacity is impaired, the appetite becomes poor; slight nausea is possible. The first complaint is often the burning sensation in the tongue. This is explained by the development of atrophic glossitis which usually attends this disease. The patient often develops achylic diarrhoea or, on the contrary, persistent constripations. Dystrophic changes in the nervous system cause skin anaesthesia and paraesthesia; the gait is oftne affectd in grave cases: spastic paresis develops (incomplete spastic paralysis of the lower extremities); the knee reflex disappears, thefunction of the urinary bladder and the rectum can also be affected. All these symptoms are known as the funicular myelosis which develops due to the predominant affection of the lateral spinal columns. Symptoms of the dosordered activity fo the central nervous sytem (dranged sleep, emotional lability, etc.) become apparent.

Inspection of the patients reveals pallor of the skin and mucosa, usually with a yellowish tint due to increased decomposition of megalocytes and formation of billirubin from the released haemoglobin, and a slight swelling of the face. The patient is not thin. Quite the reverse: most patients are well fed. The bright-red smoth and glossy tongue (because of the pronounced atrophy of the papillae) is quite characteristic of the Addison-Biermer anaemia. This symptom is known as Hunter's glossitis (W. Hunter was the first to descrie this symptom). The mouth mucosa and the posterior wall of the throat are also atrophied. The tip and edges of the tongue, and also the mouth mucosa can be ulcerated. The tedency to caries is often seen in the teeth.

Pressing or tapping on the flat and some tubular bones (especially the tibia) is often painful. This is sign of bone marrow hyperplasia. Palpation can reveal a slight enlargement of the liver and the spleen.

The cardiovascular system is usually involved as well. The left border of the heart is displaced to the left, tachycardia develops, "anaemic" systolic murmur is heard at the heart apex in 75 percent of cases; the nun's murmur is often heard over the jugular veins. The pulse is soft an accelerated. Most patients develop hypotension. ECG shows a certain decrease in the general voltage, the decreased T wave and the S-T interval.

Changes in the gastro-intestinal tract are pronounced. Especially characteristic is atrophy of gastric mucosa which can be revealed by X-ray examination,k and more distinctly by gastroscopy. The atrophy is often focal, and more distinctly by gastroscopy. The atrophy is oftne focal, and the affected sites (mostly in the fundus of the stomach) can be seen as iridescent spots. Atrophy can combine with polyps in the folds of gastric mucosa and its polypous thickening. It should be remembered that anaemia, including pernicious anaemia, can be a symptom of a malaignant tumour in the stomach. Cancer of the stomach occurs in patients with the Addison-Biermer anaemia 8 times more frequently than in healthy persons. Patients with this disease should therfore be systematically inspected by X-rays (by gastroscopy whenever possible). Almost all patients develop achlorhydria. In 98 per cent of cases it has the histamine-resistant character. The total amount of juice produced during the study is usually significantly diminished; the pepsin content of the juice low or it cannot be determined at all (achylia). Usually achlorhydria develops many years before the first symptoms of anaemia develop.

Elevated temperature is a common symptom of vitamin B_{12} (folic acid) deficiency anaemia; the temperature is usually subfebrile.

Blood plasma containsslightly increased amounts of free bilirubin due to increaed haemolysisof the red cells, especially meglocytes; the plasma iron content increased to 30-45 mmol/l (170-200 μg/ml).

The blood picture is characterized by a sharp decrease in the quantity of erythrocytes (to 0.80×10^{12} per 1 l) at a comparatively high haemoglobulin saturation. Despite the decreased total haemoglobin content of the blood, the colour index remains high (1.2– 1.5). Red blood cells differe in size (Anisocytosis), with prevalence of large erythrocytes (macrocytes). Especially large slightly oval and intensely red megalocytes appear (in many cases megaloblasts are also seen). The volume of each cell increases. Many erythrocytes are oval, or they have the shape of a sickle and other shapes (poikilocytosis). Megalocytes often have remnants of the nucleus or its envelope in the form of Jolly bodies or Cabot rings. The content of reticulocytic crisis) during vitamin B_{12} therapy to indicate the beginning remission. Blood leucocytes decrease mostly at the expense of neutrophils. Eosinopenia, relative lymphocytosis, and thrombocytopenia are observed. Large neutrophils with polysegmented nuclei also occur.

The quality of erythroid precursors in a specimen of bone marrow sharply increases, by 3–4 times compared with the number of leucopoietic cells (the proportion being reverse in physiological conditions). Megaloblasts are observed in varying amounts among the erythroid precursors; in grave cases they are found in prevailing quantity. Both erythrropoiesis and leucopoiesis are disordered. Megakaryocytes are also large, with a multi-lobed nucleus: thrombocyte separation is disordered.

Course

If untreated, the disease progresses. Before Minot and Murphy proposed their effective treatment of the disease, patients rarely survived more than 3 years. At the terminal period, many patients developed coma (coma perniciosum) with loss of consciousness, arephlexia, decreased arterial pressure and temperature, vomiting and involuntary urination.

At the present time the patient recovers from the Addison-Biermer anaemia if treated properly and if adequate prophylactic measures against relapses of the disease are taken.

Treatment

Vitamin B_{12} is given. In most cases treatment begins with moderate doses of the vitamin (100–300 μg) which is given once a day intramuscularly or subcutaneously. Considerable shifts in the bone marrow punctuate toward normalization or erythropoiesis are observed already in 24 hours after the first dose of the vitamin is given. Cells produced by division and differentiation of juvenile forms are very much like the cells produced at the corresponding stages of normla erythropoiesis. Erythropoiesis normalizes completely in 2 or 3 days. In 5 to 6 days of the therapy, thenewly formed erythrocytes enter the blood vessels in considerable amounts; the reticulocytic crisis occurs. The number of reticulocytes in the peripheral blood increases to 20-30 per cent and then gradually decreases. General weakness lessens, work capacity is regained, and gastric secretion normalizes in certain cases.

Signs of funicular myelosis are eliminated much slower and do not always disappear completely. After the blood picture normalizes and the symptoms of the disease markedly subside, the patient is given maintenance therapy with small doses of vitamin B_{12} (100 μg weekly, or 2–3 times a month). This therapy should be maintained for the rest of the patient's life. Clinical blood counts should be done periodically. The treatment ensures adequate subjective condition of the patient, his work capacity is regained, and relapses of the disease are prevented.

Auto-immune Haemolytic Anaemia

Aetiology and pathogenesis

The pathogenesis of acquired auto-immune haemollytic anaemia (anaemia haemolytica chronica) is underlain mainly by the immunopathological shifts, whichar manifested by the production of antibodies to own erythrocytes (autoagglutinins). This condition may be caused by acute infections, poisoning, medicamentous in other factors. These antibodies belong to the immunoglobulin fraction and are incomplete or "weak" antibodies. Erythrocyte-bound antibodies do not cause agglutination in the blood vessels but block erythrocytes to promote their depositionin the reticulohistiocytic system (mostly in the venous sinuses of the spleen) and their capture and destruction by macrophages.

Sometimes auto-immune haemolytic anaemia is associated with the appearance of cold antibodies which are boudn (together with the complement) to erythrocytes. Their action is manifested in the peripheral parts of the body (finger tips, ears) in overcooling. In addition to auto-agglutinins, autohaemolysins are also found in some patients. The disease may proceed in them with with signs of both extra- and intravascular haemolysis.

Clinical picture

The disease develops either gradually and incidiously or acutely with a haemolytic crisis. The main complaints of the patient are weakness, dizziness, fatigue and slightly elevated temperature. All these symptoms intensify during the haemolytic crises. Skin itching is absent. The skin is pallid with a slightly icteric hue. Applying pressure to the sternum and its percussion are painful. Palpation reveals enlarge and consolidated spleen; the liver is enlarged only slightly.

The blood erythrocyte and haemoglobin content is low, while the colou rindex remains normal. Erythrocytes vary in size, shape, and colour (poikilocytosis, anisocytosis, anisochromia). The average size of the erythrocytes is slighly smaller than normal (microcytosis). As distinct from congenital haemolytic anaemia, the erythrocytes of healthy individuals, as well as of patients with acquired anaemia, are less intensely coloured in the centre than by the periphery, which depends on their form (planocytes). The number of reticulocyts is high; reticulocytosis is especially marked in considerable anaemization and after the haemolytic crisis. The osmotic resistance of erythrocytes is not substantially changed. The blood serum of patients is yellowish. Study of the blood confirms increased content of unbound bilirubin on which the colour depends. *hypergammaglobulinaemia* is also determined. The iron content of the serum is increased. Iron is liberated in large amounts during haemolysis of erythrocytes. Owing to increased liberation of bilirubin, bile obtained by duodenal probing is very dark. The urine and faeces of the patient are darker than normal; daily excretion of stercobilin with faeces and of urobilin with urine is increased. Study of specimens of bone marrow indicates more or less significant intensification of erythropoiesis.

Both cell-bound (blocking) antibodies and those found in the free state in the plasma (conglutinins) are revealed in the blood of patients with auto-imune haemolytic anaemia. Coombs' test is used to reveal them. A direct Coomb's test is used to reveal cell-bound (blocking antibodies. A suspension of the patient's erythrocytes washed in isotonic sodium chloride solution is added to the serum of a rabbit immunized by human blood globulins. Erythrocytes agglutinate if anti-erythrocytic antibodies are present on their surface. Erythrocytes of individuals without acquired haemolytic anaemia are not agglutinated. Conglutinins in the serum of patients are revealed by addign erythrocytres of a healthy individual (donor) in order to absorb the antibodies on them. The cells are washed and carried through the test as described above. This is the indirect Coombs's test.

Course

The course of the disease is usually undulant. The disease is exacerbated by infections, big does of some medicines, e.g. salicylates, and by some other transient factors. In grave and long-standing cases, the patient's bone marrow may be exhausted and anaemia becomes hyporegeneative. In some cases the activity of the bone marrow may also be inhibited by the production of auto-antibodies to erythroblastic precursors. Formation of the pigment stones in the gall bladder is a complication of the disease. *Thrombophlebitis*, and *thrombosis* of the splenic vein are other possible complications.

Treatment

Corticosterods inhibit the production of *antierythrocytic auto-antibodies.* Blood transfusion should be carried out in rare cases because it can markedly enhacne haemolysis.

Meloplastic Syndrome (Panmyelophthisis)

The *myeloplastic syndrome* or *panmyelophthisis* is a large group of conditions of various aetiology and pathogenesis, whose main clinical symptoms are determined by the inhibition of blood formation in the bone marrow. Congenital (gentically determined) and acquired forms of *myeloid aplasia* ar distinguished by the origin, and acute and chornic forms—by the course of the disease. There are also forms characterized by incomplete inhibition of the regenerative capacity of the bone marrow (*hypoplasia*), and complete functional inhibition (aplasia).

Various clinicohaematological variants of myeloid hypo-and aplasia are differentiated by partial (in one direction) or total (in all directions) inhibition of the regenerative capacity of the bone marrow. Most specific forms are *hypo-* and *aplastic anaemia* (in which the erythropoietic function of the bone marrow is inhibited in the first instance), *agranulocytosis* (inhibition of the granulocytopoietic function of the bone mairo w), and also *panmyelophthisis* in which the regenerative function of the bone marrow is inhibited in all directions (the production of erythrocytes, granulocytes and thrombocytes is deranged more or less uniformly).

Hypo- and aplastic conditions of the bone marrow (especially *hypoplastic anaemia*) should not e mistaken for the hyporegenerative conditions which develop for example in chronic posthaemorrhagic anaemia, grave haemolytic anaemia, and in some other conditions. As sistinct from hypoplastic anaemia, a sufficiently large amount of the erythroid precursors are preserved in the bone marrow of patients with hyporegenerative anaemia; the course of the disease is nto steady; if the cause of the bone marrow exhaustion is removed, its function is restored. The myeloplastic conditions do not also include cases where the decreased content of the formed elements (erythrocytes, neutrophils, thrombocytes) in the peripheral blood is explained by their increased destruction in the spleen (hypersplenism) which occurs when the spleenis significantly enlarged (E.g. in cirrhosis of the liver) because the activity of the bone marrow is not decreased in these conditions but is, on the contrary, increased. Cytopenia due to metaplasia of the bone marrow (in malignant lymphomas and leucosis) and due to displacement of myeloproliferative tissue (in multiple myeloma, multiple metastases of cancer, etc.) does not belong to this group either.

Aetiology and pathogenesis

The aetiological factors of myeloid hypo-and aplasia are varied. These may be endogenous factors, e.g. thymus hypofunction, hereditary prpdisposition, etc. But the exogenous factors ar as a rule decisive. Among them the most important are (a) chemical poisoning, e.g. with benzene, tetrathyl lead; (b) prolonged and uncontrollable medication with some drugs, e.g. amidopyrin, butadion, cytostatic preparations (embichin, phenicol), or increased sensitivity to them; (c) infectious and toxic effects (tuberculosis, sepsis, syphilis); (d) vegetable food that may, under certain conditions, contain toxic substancs (wheathered corn, etc.); (e) ionizing radiation (radioactive substance, X-rays, etc.). It is believed that the above factors act mainly by inhibiting the nucleoproteis enzymes to slow down the emitotic cell division.

A special importance is now attached to auto-aggressive mechanis-min the development of some forms of myeloid hypoplasia. Thsi mechanism consistsin productionofantibodies to own bloodand bone marrow cells.It is possible however, that the auto-antibodies are not the triggering mechanism of the disease, and their production is only secondary to the changes in the cells. Cause of myeloid aplasia remain unknown in some cases.

Pathological anatomy

Pronounced anaemia of organs, dystrophic changes and traces of multiple haemorrhage in them are revealed in classical cases. The territory of red bone marrow is markedly contracted due to its replacement by fat tissue.

Clinical picture

The clinical picture of the disease varies depending on the prevailing direction of the inhibition of the bonr marrow function. But the most characteristic symptoms are caused by *anaemization*, *haemorrhagic diathesis, necrotization* of tissues, and secondary infection (e.g. in *agranulocytosis*). As the disease progresses, certain changes can be seen in the clinical picture: a certain syndrome may dominate during the initial stage of the disease but later other syndromes join to indicate the total inhibition of the regenerative capacity of the bone marrow. If the disease develops gradually and remains unnoticed by the patient his first complaints ar weakness, dyspnoea, rapid unmotivated fatigue, and decreased work capcity. Haemorrhagic signs are possible in thrombocytopenia: nasal bleeding, multiple ecchymoses on the skin caused by the slightest injuries and sometimes spontaneously, gastro-intestinal and uterine haemorrhages. Fever is also characteristic.

Inspection of the patient reveals pallor; signs of resolvign subcutaneous haemorrhages can be seen as bluish-purple spots, which later turn brown, and finally yellow. The skin is moist, its turgor is slightly decreased. The tourniquet and pinch tests are positive in pronounced thrombocytopenia. Changes in the heart, lungs, kidneys and the gastro-intestinal tract are usually not pahognomonic, but haemo-rrhages into various organs and internal and external haemorrhages are possible in the haemorrhagic syndrome. The lymph nodes, the liver and the spleen are usually not enlarged.

The blood picture shows various degrees and directions of changes in the regenerative capacity of the bone marrow. Hypo- and aplastic anaemia are characterized by markedly decreased erythrocyte counts (to 1 × 1012 per 1 l and below). As a rule, *erythrocytopenia* is attended by more or less pronounced leuco- and *thrombocytopenia*. The number of retuculocytes decreases considerably.

Neutropenia is most pronounced in agranulocytosis: the number of leucocytes decreases to 1.5×10^9–1×10^9 and even to 0.2×10^9 per 1 l. The number of granulocytes decreases in the first instance. Their percentage content does not exceed 15–5 of the total number of leucocytes. This accounts for relative lymphocytosis and monocytosis although the absolute content of these cells forms usually remains unaltered or decreases only insignificantly. Younger forms of neutrophil leucocytes (stab neutrophils) are practically absent from the peripheral blood. Eosinophils cannot be revealed either. Some pathological changes in the nuclei and cytoplasm of neutrophils are seen: *pyknosis* of the nuclei and toxic granulation of the cytoplasm. Anaemia and thrombocytopenia are not pronounced during the initial period but later they become more pronounced.

In certain cases, anaemia, leucopenia and thrombocytopenia develop almost simultaneously which corresponds to the clinical picture of *panmyelophthisis*. When the number of thrombocytes falls below 30×10^9 per 1 l, the haemorrhagic syndrome develops.

The *sternal punctate* is not substantially changed durign the early stages of the disease. The number of erythroid or myeloid precursors may be relatively decreased. In marked cases, the sternal punctate can contain only mearge quantity of cell elements (with prevalence of the reticular andplasma cells). Since the course of marrow aplasia is not uniform, sternal punctuate may occasionally contain portions of bone marrow in which the changes are only insignificant. This may mislead a physician in his assessments of the disease gravity. A more correct and accurate picture of marrow haemopoiesis gives *trepanobiopsy* of the iliac bone by which histological preparations of the bone marrow can be obtained.

Course

Acute, subacute, and chornic forms ar distinguished. The acute disease may have a fulminant course: cases were reported in which patients died in two days. The course of the forms with inhibited leucopoiesis is usually more rapid than that with the prevalent inhibition or erythropoiesis. This to a certain degree depends on a different life span of white and red blood cells.

The prognosis of the congenital and genuine myeloid aplasia is quite unfavourable. The disease progresses and the patient dies in a certain period of time. The progress of the myelotoxid form of the disease, e.g. caused by overdosage of amidopyrin, butadion or cytostatics, can often be arrested if the intake of these preparations is discontinued in due time. The patient may thus recover. The frequent cause of death in agranulocytois is sepsis.

Treatment

The patient must be taken to hospital. In order to act on the autoimmune mechanism of the disease, corticosteroids (e.g. prednisolone) are given. If anaemization is pronounced, repeated transfusions of blood and erythrocytic mass are indicated. Blood and specially prepared leucotyic and platelet mass are transfused to patients with leucopenia and thrombocytopenia. In order to stimulate leucucopoiesis, sodium nucleate, pentoxyl, and indicated in all cases. Antibiotics are given in septic complications. Bone marrow transplantationis now used to trat myeloid aplasia.

Haemoblastosis

Haemoblastosis is proliferation of the haemopoietic tissue; it can be diffuse and focal.

Haemoblastosis is a disease of the whole blood system charracterized by (1) progressive cell hyperplasia in the haemopoietic organs with pronounced prevalence of proliferation of certain cells (which determine the morphological essence of the disease in each particular case) over their differentiation (maturation), and the loss of their typical morphological and functional properties; (2) substitution (metaplasia) of these pathological cells for normal cells of the haemopoietic organs; (3) development of pathological foci of haemopoiesis in various organs.

Haemoblastosis is a comparatively rare disease. Its mortality rate is 1.7–8.1 per 100 000 population. The share of leucosis among therapeutic cases is 1.5–2.6 per cent. However the incidence of haemoblastosis, especially of acute forms,k has recently increased in all countries.

Aetiology and pathogenesis

Most authors regard haemoblastosis as tumours whose morphological basis are haemopoietic cells of various organs. This has been proved by numerous observatiosn and experiments. For example, some factors can provoke the growth of tumours and haemoblastosis. These factors are cancerogenic substances (3,4-benzpyrene, methylcholanthrene, etc.) and radiation. There are common features in the character of tissue proliferation in haemoblastosis and tumours. Metabolic disorders in haemoblastosis-affected cells and tumour cells have been proved to be of the same type (anaerobic glycogenolysis prevails in ministration of an extract of tumour tissue to experimental animals causes haemoblastosis in some of them; and vice versa, administration of the bone marrow or lymph node punctates of haemoblastosis patients may provoke the growth of tumours.

There are two main theories explaining the aetiology of haemoblastosis and tumours. These are the virus and the genetic theory. At the present time more than 20 viruse have been isolated that can cause haemoblastosis in animals. Attempts at isolating the virus of the main forms of haemoblastosis n man end in failure. According to the genetic theory, haemoblastosis develops due to the congenital or acquired damage to the chromosome strructures of low differentiated cells of the haemopoietic organs. A clone theory has been launched recently, according to which haemoblastosis arises due to primary chromosome mutation in one of the haemopoietic cells with its subsequent multiplication and formation of a clone of blast cells.

Nomenclature and classification of haemoblastosis

All types of haemoblastosis are designated in accordance with the name of cells which determine the cytomorphological essence of the disease. For example, acute myeloblastic leucosis, chronic erythrom-yelosis, lymphoid leucosis, etc. The traditional names of certain types of haemoblastosis describe the major syndrome of the disease, e.g. osteomyelosclerosis, macroglobulinaemic haemoblastosis, etc. Certain types of haemoblastosis have a second name (the name of the author who was the fist to study or describe the disease), e.g. Waldenstrom disease, Cesairs disease, etc. Like other tumours, haemoblastoses can be benign and malignant.

More than 30 forms of haemoblastosis have been well studied. Recent advances in haematology have made it possible to improve the classification of the disease.

The following two groups of haemoblastosis ar distinguished: leucosis and haematosarcoma (malignant lymphoma).

Leucosis is the disease of the haemopoietic cells with the primary locus of the tumour in the bone marrow. The release of tumour (leucosis) cells into the blood cause leukaemia as a symptom of the disease. The following forms of leucosis are distinguished.

1. Acute leucosis in which haemopoiesis is transformed at the expense of low differentiated blast cells or precursor cells of the 3rd and even 2nd series. Haemocytoblastosis is now absent from the modern classification because it has been proved that the "youngest" cell of haemopoiesis is not the haemocytoblast but three consecutive series of precursor cells. Various forms fo acute leucosis ar differentiated mainly by the cytochemical characteristics of leukaemic cells. In adult patients, acute myeloblastic leucosis occurs in 60 per cent and lymphoblastic leucosis in 25-30 per cent of cases. Other forms occur less frequently.

2. Chronic leucosis, in which haeopoiesis is transformed at the expense of more mature cells of differentiated haemopoiesis. Separate forms are differentiated quite easily by cytomorphological signs of leukaemic cells: chronic myeloid leucosis, chronic lymphoid leucosis, chronic erythromyelosis, erythraemia, etc.

It should be remembered that differentiation between acute and chronic leucosis first of all depends on the cytomorphological sign (the degree of cell maturity) rather than on the clinical course of the disease, although both these signs coincide in most cases. Acute leucosis may continue for a year or even longer, while a patient with chronic leucosis may die within a few months. thErefore, the diagnosis, in addition to the definition of the type of leucosis (acute, chronic), should also determine its course. It is also important to remember that acute leucosis almost never transforms into the chronic form because it is characterized by anaplasia (the cells lose ability to further maturation) of the initial elements of haemopoiesis. Quite the reverse, chronic leucosis can be exacerbated with formation of non-differentiated elements (blasts) in the foci of haemopoiesis and in the peripheral blood. Exacerbations may be spontaneous or provoked by some causes of progressive anaplasia of leukaemic cells. Chronic leucosis in the stage of blast crisis is similar to acute leucosis.

Leucosis can have the following three variants: with considerable increase in the quantity of pathological cells in the peripheral blood (*leukaemic form*), with moderate increase in their number (*subleukaemic form*), and witout appreciabale leukaemic shift in the presence of normal or decreased quantity of white blood elements (*aleukaemic form*).

Leukaemic forms of leucosis are sometimes difficutl to differentaite form leukaemoid reactions (extreme leucocytosis to 0.1×10^{12} per 1 l of blood) which are observed in some infectious diseases. In leukaemoid reactions, the leucocyte counts can be shifted to the left, to myeloblasts. But this leucocytosis is leukaemoid reactions is characterized by its potential reversibility (provided the causes are removed) and by preservation of the functional and morphological properties of leucocytes. Although leukaemic cells are called normal haemopoietic cells, in fact they cannot perform their usual function, and patients with leucosis have decreased immunity. The cells cannot differentiate into more mature cells; they decompose at a quicker rate. It has been established that leukaemic cells have an inhibiting effect on normal haemopoiesis.

Leucosis is characterized by lability, pronounced ability to transformation with exacerbation of the process (With the progress of anaplasia to less differentiated cells) and with involvement of other haemopoietic cells. Moreover, bothmorphological and clinical changeability of the process are characteristic. In some cases more or less long remissions can occur spontaneously or under certain effects (infectious disease, treatment). Remissions are then followed by exacerbations (to blast crisis).

Each type of leucosis is a static expression of the specific features of cllinico-cytomorphological picture of the diseases and at the same time can also include certain elements of the dynamic process. This should be taken into consideration during establishing a diagnosis and a prescribing the appropriate therapy.

Haematosarcoma (malignant lymphoma, or regional tumours with their possible generalization) is also a tumour of haemopoietic cells but its localization is mostly extra-marrow and focal. Lymphogranulomatosis is the most common haemoblastosis. Paraproteinaemic haemoblatosis (multiple myeloma, Waldenstrom's disease) is a special form arising from precursors of B lymphocyes and characterized by production of pathological globulins. Reticulosarcooma and some other diseases are also special forms of paraproteinaemic haemoblastosis.

Leucosis can also transform into some types of malignant lymphomas, except lymphogranulomatosis. The reverse process is also possible: ample leukaemic cells appear in the blood and bone marrow in malignant lymphomas, i.e. the blood picture and the course of the diseases are similar to those observed in leucosis.

Acute Leucosis

Acute leucosis is characterized by profuse proliferation of the youngest (blast) elements of blood with their subsequent disturbed differentiation and also with development of foci of pathological haemopoiesis in various organs. Lympho-and myeloblatic forms of the disease are common.

Acute leucosis occurs at any age, but men and women from 20 to 30 are mostly affeced.

Pathological anatomy

The skin and the internal organs are anamic. Mucosa (especially in the fauces and mouth) and often the skin ar affected by necrotic ulcers and multiple haemorhages. Haemorrhages into th einternal organs also occur.Lymphnodes, the spleen and the liver are moderately enlargaed. Histological studies of these organs and of the bone marrow show that they are as if "stuffed" with non-differentiated cells of blood (blasts).

Clinical picture

The onset of the disease is in most cases acute or subacute: high temperature (remittent or hectic), profuse sweating, chills, pronounced weakness, pain in the bones, and other general symptoms resembling those of acute septic affections. Pain in the throat of often one of the first complaints: swallowing becomes painful because of necrotic ulceration of the thorat and fauces. For this reason the disease is often mistaken forr necrotic tonsillitis and only further observation of the patient and the study of the bone marrow and blood help the physician establish a correct diagnosis. Fever, chills and sweating, which ar so characteristic of acute leucosis, are explained by the pyrogenic effect of purines released in great quantity during the decompositon of immature leucocytes. Fever can also be caused by secondary infections; despite the markedly increased production of white blood cells, their function is inadequate, and resistance of leucosis patients to various infections decreases.

In some cases the onset of the disease is gradual, with non-pronounced general symptoms: slight weakness, indisposition, rapid fatigue, and subfebrility. The condition then worsens and a complete clinical picture of the disease develops. Anaemia and various haemorrhagic complications and secondary infection develop.

Inspection of the patient usually reveals grave condition from the very onset of the disease. In the terminal period the condition is particularly grave: the patient is passive, answers the doctor's questions with difficulty, or is unconsious. The skinis pallid, sometimes with yellowishor greyish hue, and moist; its turgor is decreased. Traces of subcutaneous and intracutaneous haemorrhages can be seen. The tourniquet and the pinch tests are positive; haemorrhage is considerable at points of injections. Necrosis and bed-sores are possible. Necrosis of the mucosa, especially of the mouth and throat, is especially pronounced. Ulcerous and necrotic tonsillitis, gingivitis, and somatitis are quite characteristic of the disease. Necrotized surfaces are covered with a poorly removable grey or yellowish coat. When removed, it reveals bleeding ulcers. The breath of the patient is putrefactive. Palpation reveals enlargement of separate groups of the lymph nodes, spleen, and the liver. The heart borders are broadened; tachycardia, systolic murmur at the heart apex (due to dystrophic processes in the heart muscle), and anaemia are revealed. Pericarditis and pleuritis are possible. Each haematological form of leucosis is characterized (though not necessarily) by some special clinical features.

The blood of patients contains increased number of white blood cells: to 1×10^{11} and even 2×10^{11} per 1 l (in rare cases even this figure may be exceeded) (Plate 32). Subleukaemic forms of the disease can occur. Leucopenia can develop in some cases at the early stage of acute leucosis. Leucopenia is then succeeded by leucocytosis. The most specific haematological sign of the disease is the presence of blast cells in the peripheral blood. All last cells are similar morphologically but special cytochemical reactions can be used to differentiate between them. Prevalence of their certain forms is determined by the haematological variant of leucosis (acute lymphoblastic, acute myeloblastic, acute monoblastic leucosis, etc.). The immature forms may amount to as high as 95 and even 99 per cent. Leukaemic cells often have some specific defects in their nuclear and cytoplasmic structure. Only the youngest and the most mature cells can be revealed in the blood of most patients with acute leucosis, while intermediate forms ar absent (hiatus leucaemicus). Eosinophils are basophils are absent; other cell forms are decreased significantly not only relatively but also absolutely. Thrombocytopenia and anaemia are observed which can be explained by the displacement of megakariocytes and erythroblasts from the bone marrow by vigorously proliferating blast cells and also by the prevalent development in the direction of leucopoiesis. Anaemia may intensify due to haemorrhage (characteristic of this disease) and also due to the intensified haemolysis of erythrocytes. Coagulability of blood and the bleeding time are abnormal in most cases: ESR sharply increases.

The bone marrow punctate contains 80-90 per cent of leukaemic blast cells, which displace all other cell elements.

Course

The course of the disease is progressive. Prognosis is unfavourable. The average life expectancy of patients with acute leucosis is about 2 months; in separate cases from 2 days to 18 months. But modern therapy can prolong the patients' life to 2–3 years; in rare cases to 5 years and more.

Treatment

Combined therapy, including 3–5 preparations, depends on the clinico- haematological variant of the disease. Corticosteroids (e.g. prrednisolone) in large doses are prescribed together with cytostatic (6-mercaptopurin, vincristin, methotrexate, etc.). Anaemia is removed by blood transfusion and by preparations preventing haemorrhagic complications (vicasol, calcium chloride, aminocaproic acid). If secondary infection arises, antibiotics are indicated. Intense vitamin therapy is also useful.

Chronic Myeloleucosis

Chronic myeloleucosis is the most common variety of leucosis. It develops from precursors of myelopoiesis. It is characterized by myeloid hyperplasia of the bone marrow attended by delayed maturation of the myeloid elements at a certain stage of their development and by myeloid metaplasia of the spleen, liver, lymph nodes and other organs. Karyological studies reveal, in the overwhelming majority of cases, the so-called Philadelphia (Ph^1) chromosome in the myeloid precursors. Chronic myeloleucosis occurs at any age but its incidence at the age of 20–45 is higher.

Pathological anatomy

The internal organs ar pallid and anaemic. The spleen is markedly enlarged, consolidated; it has traces of past ischaemic infarctions and new infarctions. Microscopy does not reveal spleen follicles; diffuse proliferation of the myeloid tissue can be seen. The liver is enlarged and myeloid proliferatoin can be traced by the course of the liver capillaries and the periportal layers. The lymph nodes are somewhat enlarged; their section is greyish0red. Myeloid tissue can be seen in them (as well as in the other organs).

The bone marrow is juicy, bright-red or greyish-red; it displaces fat marrow) in the tubular bones to a lesser or greater extent. Red bone marrow is represented mainly by the myeloid elements; the more acute the process, the greater the prevalence of less differentiated elements.

Clinical picture

The initial symptoms of the disease are not specific: weakness, fatigue, excess sweating, and subfebrile temperature. The symptoms become more pronounced with time and the patient consults a doctor. The work capacity decreases appreciably, weakness increases, sweating becomes profuse, the body temperature rises periodically to 37.5–39°C; cachexia develops. A common symptom is the feeling of heaviness in the left part of the abdomen whichdepends on the pronoucned enlargement of the spleen. Pain may be due to a considerable distension of the spleen capsule. Splenic infarction is manifested by piercing pain which intensifies during breathing. Pain in the bones is not infrequent. It is due to hyperplasia of the myeloid tissue.

Myeloid infiltration in various internal organs can be the cause of some additional symptoms, such as dyspeptic signs in affections of the gastro-intestinal tract, coughing in the presence of infiltrations in the lungs and the pleura, neurological changes due to affection of the brain, the spinal cord, nerve radices, etc. In the terminal period of the disease, the heart is overloaded due to pronounced anaemization; dyspnoea and oedema develop (their origin being dependent also on hypoproteinaemia). Thrombocytopenia and shifts in the blood coagulation system cause haemorrhagic complications. Inspection of the patient helps physician assess his general condition and determine (approximately) the stage of the disease (stage I–the initial symptoms, stage II–pronounced symptoms, and stage III–dystrophy; this is the terminal stage). The terminal stage is characterized by pronounced cachexia and considerable enlargement of the abdoomen due to markedly enlarged liver and spleen. The skin is pallid, with a yellowish or greyish hue; it is flaccid and moist. The legs are affected by oedema. Gingivitis and necrosis of the mouth mucosa are possible. Palpation reveals moderate enlargement of the lymph nodes of various groups. The liver and especially the spleen are markedly enlarged. It is believed that no other disease is attended by enlargement of the spleen to the extent to which it is enlarged in the terminal stage of chronic myeloleucosis. The liver and the spleen are firm. In the presence of infarctions of the spleen it is tender to palpation. Peritoneal friction sound can be heard over the spleen by ausculation. Applying pressure to the bones and tapping over them are painful.

The leucocyte count is markedly high (Plate 33). It can be 3×10^{11} and even 6×10^{11} per 1 l. Leucocyte count can first only slightly exceed the normal level but later it increases, gradually or suddenly.

Temporary remissions are possible, especially in the appropriate therapy. In addition to the main form of myeloleucosis characterized by a considerable increase in the white blood cells count (leukaemic), there may be cases with their moderately increased (subleukaemic) and normal counts (aleukaemic). Study of the blood smears reveals mainly cells of granulocytic series which make 95-97 per cent of all white blood elements. There are many immature forms among them (myelocytes, promyelocytes, and even myeloblasts). During exacerbation, the number of young forms markedly increases. Only the youngest cells, myeloblasts, and a relatively small number of mature granulocytes (stab and segmented ones) can be revealed in the blood, while th eintermediate forrms are absent (hiatus leucaemicus). Basophils and eosinophils are usually present in the smear; their percentage can even be increased. Basophilia in 4–5 per cent of cases is regarded as a sign of myeloleucosis. The number of lymphocytes and monocytes decreases to 3–0.5 per cent in grave cases with singificant leucocytosis, but their absolute amount in the blood does not change substantially. Red blood changes are only observed at stages II and III of the disease, when anaemia joins the process and progress. Erythrocyte and haemoglobin content of the blood decreases synchronously. The colour index therefore remains within the normal range (0.8–1.0). Thrombocytopenia develops as the terminal period approaches. ESR usually increases to 30–70 mm/h.

The content of the erythroid precursors markedly decreases in the bone marrow (especially when the disease approaches its terminal stage). Cells of the myeloid series perevail (especially juvenile forms: promyelocytes, myelocytes, and myeloblats). Megakaryocyte coutn slightly increases in the first half of the disease. Characteristic also is the increase in the number of basophilic and eosinophilic promyelocytes and myelocytes.

Course

The course of the disease is progressive, sometimes with transient spontaneous remissions. Before modern methods of treatment of the disease were introduced into clinical practice, the average life expectancy of patients was 2.5–3 years (sometimes to 10 years). Today the life of patients is prolonged more significantly. Patients die of cachexia, anaemization incompatible with life, haemorrhagic complications, or from a joining infection.

Treatment

Myelosan and other cytostatics are prescribed, usually in

combination with 6-mercaptopurine, prednisolene, etc. Patietns with a marked splenomegaly are given radiotherapy or dopan. Repeated transfusion of blood or packed red cells are indicated in cases with pronounced anaemization.

Chronic Lymphoid Leucosis

Chronic lymphoid leucosis is now regarded as a benign tumour of the immunocompetent tissue. Its haematological basis is mainly B-lymphocytes (morphologically mature but functionally inadequate). Chronic lymphoid leucosisis characterized by systemic hyperplasia of the lymphoid apparatus, lymphoid metaplasia of the spleen, bone marrow, and other organs. Chronic lymphoid leucosis is a common form of lecosis. It usually occurs in the middle-aged and aged individuals (from 35 to 70), mostly in men.

Pathological anatomy

The lymph nodes of various groups are enlarged singificantly. Their section is grey or greyish-red. The pattenr of the lymph nodes is blurred; microscopy reveals accumulationof lymphoid cells, among which juvenile forms occur. The tonsils are enlarged, their structure is indistinct; microscopy reveals large accumulations of lymphocytes and lymphopoietic cells. The spleen is markedly enlarged but not to the extent to which it is enlarged in chronic myeloleucosis. Its structurre is indistinct because of diffuse hyperplasia of the lymphoid tissue. Lymphatic infiltrations are seen in the liver, the stomach wall, pancreas, kidneys, and the skin; in other words, all organs can be affected. Lymphoid metaplasia of the bone marrow is observed.

Clinical picture

The initial symptoms are general weakness, indisposition, and rapid fatigue. The first symptom which troubles the patient and makes him feel like consulting a doctor is usually enlargement of the subcutaneous lymph nodes. General weakness gradually increases, excess sweating develops along with subfebrile temperature. Depending on a particular enlarged lymph node group and the organ affected by lymphoid infiltratio, additional symptoms develop: dyspepsia, diarrhoea (in affection of the gastro-intestinal tract), dyspnoea and attacks of asphyxia (in compression of the trachea and bronchi by the bifurcation lymph nodes), erythema, dryness and itching of the skin (in leukaemic lymphodermia), etc. Lymphoid metaplasia of the bone marrow may cause haemorrhagic symptoms (due to thrombocytopenia) and anaemia. Leukaemic infiltration can cause radicular pain and exophthalmos. Diffuse lymphatic proliferation in the nasopharynx can develop.

Enlarged lymph nodes can often be revealed during inspection of the patient. Lymph nodes of one or several groups are often enlarge; later all other lymph nodes become also involved. The tonsils can be enlarged too. Skin infiltration is attended by its consolidation, reddening, dryness and scaling. At the terminal stage, the patients are extremely thin (*cachexia*).

Palpation is used to assess accurately the enlargement of the lymph nodes and their properties. The lymph nodes are elastic-pasty; they do not fuse with the skin or with one another, and are painless in most cases. They can grow to the size of a hen egg. Even markedly enlarged, the lymph nodes never ulcerate or suppurate (As distinct from tuberculosis affection of the nodes). Theliver and the spleen are enlarged and consolidated. Infarctions of the spleen can occur; its palpation then becomes tender.

Leucocyte counts in the leukaemic form of the disease are as high as 3×10^{11} per 11, and more. Lymphocytes make 80–95 per cent of the white blood; they are mostly mature. The structure of their nucleus and cytoplasm is sometimes quite peculiar; the cells are very soft and are easily destroyed when preparing a smear; specific Botkin-Gumprecht shadows are formed. Small amounts of juvenile cells, prolymphocytes and lymphoblasts, occur. During exacerbation their number increases. The relative quantity of neutrohils is much decreased (to 20–4 per cent). The blood picture is less specific in subleukaemic and aleukaemic forms of the disease, where lymphocytosis is usually less pronounced. *Anaemia* and *thrombocytopenia* (mainly of the auto-immune genesis) joint at the terminal period.

Study of the punctuate of the bone marrow reveals its lymphoid metaplasia: great quantity of lymphoid cells are found (to 50 and even 90 per cent in especially grave cases). The number of cell elements of the granulocytic and erythroid precursors is decreased. The results of studying the lymph node punctate are less convincing because lymphoid cells are common elements of its parenchyma while the hyperplastic character of the lymphoid tissue is sometimes difficult to determine.

Course

The disease progresses in cycles of gradually. The average life expectancy of patients is 4–5 yars. Some survive for 10–12 years and over. The patients die of secondary infections, usually pneumonia (which is promoted by inhibition of humoral immunity), of haemorrhagic complications, and cachexia.

Treatment

Active treatment is not given during the initial period of the disease (like in *chronic myeloleucosis*). Special attention is given to normalization of conditions of work and rest of the patients, who must be fed an adequate diet rich in vitamins and proteins. The patient should walk in the open air. X-ray therapy is given in the presence of toxicosis or if the disease is rapidly progressing. The enlarged lymph nodes and the spleen are irradiated. Leukeran, prednisolone, cyclophosphane and other chemical cytostatics are given. Blood transfusion is indicated in anaemia and thrombocytopenia.

Erythraemia

Erythraemia (chronic erythromyelosis, *Vaquez' disease*) is a benign *myeloproliferative* disease characterized by total hyperplasia of the bone marrow cell elements, which is more pronounced in the erythroid precursor. Erythraemia was first described by the French clinicist Luis Vaquz in 1892. Aged males are mostly affected.

Erythraemia should be differentiated from erythrocytosis which may be a symptom of some other diseases (chronic diseases of the lungs and the heart, essential hypertension, some kidney diseases, etc.) and erythrocytosis developing in the presence of hypoxia (at high altitudes, e.g. in highland; less frequently in pilots). The symptoms of the main disease prevail in the clinical picture of symptomatic erythrocytosis. During remission, or in cases of recovery, the erythrocyte count in the peripheral blood normalizes. Symptomatic erythrocytosis is not attended by neutrophilic leucocytosis or thrombocytosis, or else splenomegaly, which are common in erythraemia.

Pathological anatomy

Prounced hyperaemia of the organs and the presence in them of old and new haemorrhages are characteristic. Red bone marrow is affected by hyperplasia; it displaces fat from the diaphysis of the tubular bones. Histological studies reveal increased number of the erythroid series cells. The liver and the spleen are mildly enlarged and plethoric. The left ventricle of the heart is often hypertrophied (in arterial hypertension).

Clinical picture

The onset of the disease is slow and indistinct. When the symptoms are activated, the patients complainsof haeadache, heaviness in the head and noise in the ears, exertional dyspnoea, impaired memory, and skin itching. Vision and hearing function are also impaired in some cases. Pain in the abdomen can develop, probably due to excess blood delivery to the internal organs. Unbearable and burning pain can transiently develop in the finger tips (*eythromelalgia*), which can be explained by transient vascular spasms.

Inspection of the patient reveals peculiar *plethoric redness* to the exposed skin (face, neck, hands). The tongue and the lips are bluish-red, the eye conjunctiva is hyperaemic. This peculiar colour of the skin and mucosa is due to overfilling of the surface vessels with blood and its slow movement in the vessels. The greater part of haemoglobin is thus converted into its reduced form. Palpation reveals moderately enlarged spleen and live. Tapping over flat bones and exerting pressure on them are painful, which is characteristic of bone marrow hyperplasia.

Arterial pressure (both systolic and diastolic) is often increased. It is believed that arterial hypertension is a compensatory reaction of the vessels to the increased viscosity of blood. In these cases palpation of the apex beat and electrocardiography reveals left-ventricular hypertrophy.

The erythrocyte count increase and is usually from 6×10^{12} to 8×10^{12} per 1 l, and more; haemoglobin increases to 180–220 g/l; the colour index is less than 1. The total blood circulatory volume increases significantly (1.5–2.5 times) mainly at the expense of increased erythrocyte count. The blood reticulocyte content increases to 15–20% which indicates intensified regeneration of erythrocytes. Polychromasia of erythrocyte is observed. The number of eosinophils, less frequently of basophils, increases. The number of thrombocytes increases sharply to 1.5×10^{12}–2×10^{12} per 1 l of blood. ESR is slow, blood viscosity increases significantly, while coagulability of blood and the bleeding time remain normal.

Histological study of the bone marrow (obtained by trepanobiopsy) shows significantly increased number of cells of the erythroid series. The number of juvenile cells of the granulocytic series and of megakaryocytes is also increased.

Course

The disease progresses slowly. The average life expectancy of patients is 10–14 years. The outcome of the disease is myelofibrosis with progressive hypoplastic anaemia or transformation into myeloleucosis. Most frequent complications are thrombosis of the cerebral vessels, spleen, lower extremities, and less frequently of other parts of the body. Tendency to bleeding also develops which is due to both functional inadequacy of thrombocytes and the low relative content of fibrinogen in the blood. Errythraemia patients often develop gastroduodenal ulcer.

Treatment

Radioactive phosphorus (32P) is the best therapeutic means for eryrhraemia. It produces a cystostatic effect on haemopoiesis in the bone marrow. Clinical remission, lasting from 1 to 3 years is attained in most cases. New cytostatics (myelosan and marcofan) are also used. Symptomatic therapy includes phlebotomy (500 ml) at 5–7 day intervals.

Lymphogranulomatosis

Lymphogranulomatosis is a systemic disease in the group of malignant lymphomas and is characterized by the specific malignant affection of the lymph nodes, the spleen, and other organs. The disease was first described by Thomas Hodgkin in 1832. Hence another name, Hodgkin's disease.

Aetiology and pathogenesis

It is believed that the disease is underlain by *lymphogenic metastasis* from the primary focus, which obeys the law of tumour propagation. *Lymphogranulomatosis* is possibly a tumour of histiocytes in which the chromosome set is discorded (*Melburn chromosome*).

Pathological anatomy

The main changes are enlargement of various lymph nodes. Their histological studies reveal focal three-stage growths of the granuloma cells. The stage of diffuse hyperplasia of the lymph node is followed by focal, and then diffuse proliferationof the reticular and endothelial cells, and also of the cells of juvenile connective tissue. The presence of the Berezovsky-Sternberg cells is especially characteristic. These giant cells have 1, 2 or 3 large nuclei with large nucleoli. Fibrosis develops at the third stage. Necrotic foci are sometimes observed. Lymphogranulomatosis is characterized by a varied histological picture and simultaneous presence of various phases of lymphogranulomatous growths in various lymph nodes. The spleen is enlarged and firm; the secretion shows many light-grey foci of lymphogranulomatous proliferations which account for the porphyritic texture. Proliferation of lymphogranulomatous tissue can be seen in other organs, e.g. in the stomach, intestine, liver, etc.

Clinical picture

Weakness and geneal indisposition are in most cases the first appreciable signs of the disease. Skin itching is among them. It can be excruciating and the patient scratches the skin. The temperature elevates and hidrosis develops. Attacks of fever are first infrequent, but later become more constant. The temperature curve has a specific pattern, to reach 38–39.5°C at its peaks. The difference between the morning and evening temperatures is 1–2°C. the main symptom for which the patient would first consult the doctor is gradual swelling of some part of the body, mostly of the neck (due to growth of the lymph nodes). Both surface and deep-seated lymph nodes become enlarged. Considerable enlargement of the lymph nodes of any internal group can involve some additional changes in the patient's condition and cause complaints. For example, when the lymph nodes grow to compress the mediastinum, the trachea or bronchi the patient develops dyspnoea, cough, pain and the pressing sensation in the chest. If the recurrent nere is compressed, the corresponding vocal cord is affected by paresis and voice becomes hoarse. Lymphogranulomatous affectionof the stomach causes symptoms of dyspepsia (Abdominal pain, regurgitation, vomiting, etc.). If the abdominal lymph nodes and the intestine are involved, persistent constipation may develop.

Findings of physical examination show enlargement of the lymph nodes which is the most characteristic sign of the disease. The cervical lymph nodes (mostly on the right side of the neck) become enlarged in more than 50 per cent cases; the nodes of the other side of the neck become involved later. Next involved are submandibular, supra- and subclavicula, axillary, femoral, and inguinal lymph nodes; less frequently the lymph nodes of the other regiosn (occipital, ulnar, etc.) are involved.

Newly involved nodes are soft, while the old ones become very firm. They fuse together to form conglomerates but do not adhere to the skin. The nodes are painless, they do not suppurate or open, as distinct from *actinomycosis*- or *tuberculosis*-affected nodes (*scrofuloderma*). If a group of lymph nodes is enlarged considerably, the overlying skin becomes darked and then bronze due to disordered circulation; it does not however get thin.

Inspection of the patient reveals some symptoms caused by compression of the vessels and nervous trunks by the enlarged lymph nodes (regional cyanosis, dilation of the veins, Horner's symptom, etc.). Abdominal lymph nodes can be palpated if their enlargement is considerable. A tuberous tumour can be felt in the mesogastric region round the umbilicus. The spleen is enlarged and firm; it is tender in the presence of perisplenitis. Enlargement of the liver is less characteristic. Enlargement of the mediastinal lymph nodes can best of all be revealed by X-rays.

Hypochromic anaemia and *neutrophilic leucocytosis* are usually found; lymphocyte counts (absolute and relative quantities) are low. Eosinophilia stage it is 50–70 mm/h. Study of the punctate of the bone marrow is not quite informative but the Berezovsky-Strrenberg cells may often be observed.

Histological study of lymph node biopates reliably confirms the diagnosis of *lymphogranulomatosis.* The cell composition of the lymphogranulomatous tissue is as a rule quite varied but its most specific element are granulomas with giant cells to 30–80 μm in diameter (Berezovsky-Sternberg polynuclear cells which do not occur in other diseases). Study of the content of lymphnodes is more accessible but less informative.

Course

The disease is progressive; remissions are followed by exacerbations. Anaemia and cachexia develop gradually. The prevalent affection of a particular group of lymph nodes accounts for the clinical picture of the disease. In some cases compression of a vital organ can acclerate the disease and its outcome. The patient may also die of a secondary infection. The average life expectancy of the patients is from 3 to 4 years; some patients however survive for 6–8 years and more.

Treatment

Radiotherapy is quite effective. It can give remissions lasting as long as several months (after irradiation of the enlarged lymph nodes). Chemical cytostatics (usually their combinations) are used.

Haemorrhagic Diathesis

Haemorrhagic diathesis is the disease characterized by the tendency to bleeding and repeated haemorrhages; they may occur spontaneously and may be caused by injuries; injury can be quite insignificant, which otherwise would never provoke bleeding in a normal individual.

Aetiology and pathogenesis

These are quite varied. Some types of haemorrhagic diathesis are hereditary but many of them can be caused by some external factors.

Avitaminosis (deficit of vitamins C and P) is an especially predisposing factor. Some infections (long-standing sepsis, louse-born typhus, virus haemorrhagic fevers, *icterohaemorrhagic leptospirosis*), allergic conditions, some disease of the liver, kidneys, and of the blood system can also provoke the onset of haemorrhagic diathesis.

Haemorrhagic diathesis can be classified by the pathogenesis into two major groups: (1) haemorrhagic diathesis due to disordered capillary permeability (*haemorrhagic vasculitis*, vitamin C deficiency, some infectious diseases, trophic disorders, etc.); (2) *haemorrhagic diathesis* due to disorders in the blood coagulation and anticoagulation system. The latter group is further subdivided into the following conditions:

A. Haemorrhagic diathesis caused by disordered blood coagulation system:

(1) First phase: congenital deficit of plasma components of thromboplatelet formation (factors VIII, IX, XI), haemophilias A, B, C, etc.; deficit of thrombocytic components (thrombocytopathy, e.g. thrombocytopenic purpura; see below);

(2) second phase: deficit of plasma component of thrombin formation—factors II, V, X, the presence of antagonists to them and of their thrombocytopenic purpura);

(3) third phase: deficit of plasma components of fibrin formation—factors I (fibrinogen) and XII.

B. Haemorrhagic diathesis caused by disseminated intravascular coagulation (thrombohaemorrhagic syndrome or coagulopathy of consumption) in which all procoagulants are utilized during massive intravascular coagulation and the fibrinolysis system is activated.

C. Haemorrhagic diathesis caused by disseminated intravascular coagulation (thrombohaemorrhagic syndrome or coagulopathy of consumption) in which all procoagulants are utilized during massive intravascular coagulation and the fibrinolysis system is activated.

This concise classification of haemorrhagic diathesis is only conventional because several pathogenic factors are often involved. This classification covers a very large group of diseases, both hereditary and acquired, and also secondary syndromes arising against the background of the main disease (metastasizing malignant tumour, burn disease, etc.).

Clinical picture

The general clinico-morphological symptoms of haemorrhagic diathesis are haemorrhages into various organs and tissue, external and internal haemorrhages (from the gastro-intestinal tract, lungs, uterus, kidneys, etc.) and *secondary anaemization*. The disease is complicated by dysfunction of the haemorrhage-affected organs, by *hemiparesis* in disordered cerebral circulation, regional paralysis and paresis in compression of large nervous trunks by *haematomas*, *haemarthrosis* in repeated haemorrhages into the joints, etc.

Despite the great variety of haemorrhagic diatheses and certain diagnostic difficulties, accurate diagnosis is quite important for efficacious therapy in each particular case. The aetiological and pathogenic factors of the disease should be properly considered for an accurate diagnosis. *Haemorrhagic diathesis* is the subject of special study of senior medical students. Here only acquainted the reader will be in general with thrombocytopenic purpura (*Werlhof's disease*).

The *prophylaxis* of hereditary (familial) haemorrhagic diathesis includes medico-genetics studies that may give the wife and husband the necessary information and advice concerning possible complications in their offspring; if haemorrhagic diathesis is not hereditary but acquired, measures should be taken to preclude development of diseases that may promote the onset of haemorrhagic diathesis.

Thrombocytopenic Purpura (Werlhof's Disease)

Thrombocytopenic purpura is a haemorrhagic diathesis due to the deficit of blood platelets. The disease was first described by Paul Werlhof in 1735. Thrombocytopenic purpura occurs mostly in young females.

The *aetiology and pathogenesis* of the disease are unknown. It has only been established that the immune-allergic mechanism is positively involved in about 50 per cent cases: anti-thrombocytic antibodies are produced and fixed on the surface of thrombocytes to damage them and to prevent their normal separation from megakaryocytes. The triggering factors (the impetus to production of auto-antibodies) may be infection, toxicosis, individual hypersensitivity to certain foods and medicines. In some cases the disease is caused by hereditary insufficiency of certain enzyme thrombocyte systems which is probably activated by some additional factors.

Pathological picture

Multiple haemorrhage in the skin and the internal organs are characteristic. The spleen may be considerably enlarged. Separation of thrombocytes from megakaryocytes in the bone marrow is disordered (according to histological findings).

Clinical picture

The main symptom of the disease is the appearance on the skin and mucosa of multiple haemorrhages in the form of small dots (*pectechiae*) or large spots (*ecchymoses*). Haemorrhage may be spontaneous and due to insignificant injuries, mild contusion, pressure on the skin, etc. Haemorrhagic lesions ar first purple, then they darken to cherry-red and brown, and then lighten to yellow and disappear in several days. But new lesions develop to succeed the disappearing ones. Bleeding from the nose, gastro-intestinal tract, kidneys or uterus are not infrequent; haemorrhages into the internal organs (brain, fundus oculi, myocardium, etc.) are also possible. Grave and prolonged bleedings arise in extraction of teeth or in other minor operations. The torniquest test (And especially the pinch test) are positive. The spleen and the lymph nodes are usually not enlarged; tapping on the bones is painless.

Thrombocyte counts are usually less than 50×10^9 per 1 l; in some cases only single blood platelets can be found in preparations. The degree of bleeding can be assessed by the degree of thrombocytopenia. Hypochromic anaemia can develop afte profuse bleeding. The clotting timeis normal in most cases, but it can be slightly longer (due to the deficit of thromboplastic factor III of blood plateletes). The bleeding time increases to 15–20 min and more; clot retraction is disordered.Thromboelastography reveals greatly increased reaction and clotting time.

Course

Both acute and chronic recurrent forms of the disease are observed. The patient dies of profuse bleeding and haemorrhages into the vital organs.

Treatment

Removal of the spleen is indicated in grave cases: the number of thrombocytes increases in the blood of a patient and haemorrhage stops in a new days following the operation. The effect of splenectomy is probably explained by decreaed decomposition of blood platelets in the spleen and by the removal of inhibiting effect that the spleen has on thrombocytopoiesis. Blood transfusion is useful for haemostasis and blood substitution.Repeated transfusion of thrombocytic mass gives positive haemostatic effect. Vitamin P, vitamin C, calcium chloride and vicasol are given to strenghthen the vascular walls. Since the allergic factor is involved in the pathogenesis of the disease, corticosteroid hormones are quite effective in certain cases.

ENDOCRINE SYSTEM AND METABOLISM

SPECIAL PATHOLOGY

Diffuse Toxic Goitre

Diffuse toxic toxic (thyrotoxicosis, Basedow's disease) is caused by thyroid hyperfunction. The disease most commonly occurs in women between the ages of 30 and 50; the incidence in men is 5–10 times lower.

Aetiology and pathogenesis

Psychic trauma, infection (Tonsillitis, rheumatism, etc.), dysfunction of other endocrine glands (pituitary) are important for the development of the disease. Familial factors are also important: toxic goitre can often be found in close relatives.

Secretion of hormones by the thyroid gland is intensified in stimulation of hypothalamic centres which stimulate secretion of the thyrotropic hormone by the anterior pituitary lobe. The hypothalamic centres can be stimulated by various factors, by psychic traumas in the first instance. Investigations of V. Baranov and other authors demonstrate the essential role of the central nervous system in the pathogenesis of diffuse toxic goitre. The authors have proved that in many patients the development of toxic goitre was preceded by neurocirculatory dystonia which interferes with ^{131}I capture by the thyroid gland. This form of neurosis is now given great significance in the pathogenesis of diffuse toxic goitre and is regarded as precursor of this disease.

Hyperthryoidism causes changes in various tissues and organs and disturbs various types of metabolism: protein-carbohydrate, fat, mineral, water metabolism, etc. Upset function of the sympathico-adrenal system is also a very important factor which accounts for many symptoms of the disease. The role of the pituitary gland in the pathogenesis of thyrotoxicosis can not be ruled out completely because patients with thyrotoxic goitre suffer from expothalmos, while the exophthalmic factor is secreted by the pituitary gland.

Pathological anatomy

The thyroid gland enlarges uniformly or by focal hyperplasia; hence diffuse or nodular goitre. Microscopy shows intense blood filling in the thyroid gland and reconstruction of fillicular epithelium into columnar or polymorphous epithelium. Sometimes the affected thyroid gland differs only insignificantly from the normal one by the character of its epithelim and follicles; the follicles may only have cyst-like dilatations and contain little colloidal substance. Lymphocytes are accumulated and lymphoid follicles are formed.

Clinical picture

The onset of the disease may be acute or gradual, with slow development of the symptoms. The main signs of the disease are enlargement of the thyroid gland, ocular sings, and heart palpitation. The patients complain of increased psychic excitability, non-motivated anxiety, deranged sleep, hyperhidrosis, tremor of the fingers or in the entire body, frequent defaecation, wasting, and muscular weakness.

Inspection of the patient immediately reveals the special features in his behaviour; fussiness, hasty speech; sometimes the patient drops the subject quite unexpectedly and starts discussing another subject. Ophthalmopathy and some other ocular symptoms suggest hyperthyroidism. Despite preserved or even increased appetite, the patient may lose much of his weight (to cachexia). The patient's skin is smooth, warm and moist to the touch. Some patients develop diffuse pigmentation of the skin which however does not colour the mucosa. The pigment is sometimes deposited selectively in the skin of the eyelids. The hair of the head becomes thin and soft.

During inspection special attention should be paid to the size of the thyroid gland and symmetry of its enlargement. If the thyroid gland is enlarged significantly, the patient's breathing becomes stridorous. Inspection of the patient should be followed by palpation of the thyroid gland. Five degrees of thyroid enlargement are distinguished: I—enlarged thyroid gland is difficult to palpate,; II—enlarged thyroid gland is clearly seen during swallowing; III—clearly visible thickening of the neck due to goitre; IV—marked goitre; V—large goitre. Enlargement of the second and third degree occurs most frequently.

Ocular symptoms

A common symptom of diffuse toxic goitre is bilateral of the eye slits which gives an expression of astonishment to the patient's face. Another frequent manifestation is Graefe's sign: a white strip of sclera between the edge of the eyelid and the upper margin of the cornea which appears as the eyeball moves downward. Among other symptoms are Atellwag's sign (infrequent blinking), Kocher's sign (exposure of the sclera between the lower edge of the upper eyelid and the upper edge of the iris when the eyes are fixed on an upwardly moving object), and the exophthalmic symptom (protruded eyeballs). The protrusion is usually more or less uniform but asymmetry is also possible. One eye can only be involved in some cases. In grave exophthalmic goitre, keratitis, ulcers, of the cornea can also develop and the patient's power of vision can thus be endangered. The eyelids can swell, and weakness of convergence can be observed; the eyeball can move aside when attention is associated with upset function of the oculomotor muscles.

Cardiovascular system

Tachycardia is one of the most frequent symptoms of the disease. Pulse rate varies within the range of 90 to 120 and in grave cases to 150 beats per minute. Systolic and minute volumes, the mass of the circulating blood and the rate of the blood flow increase, systolic pressure grows, diastolic pressure falls, and the pulse pressure increases. Ausculation of the heart reveals a snapping first sound an dsystolic murmur at the apex and over the pulmonary artery which are due to increased blood flow rate and low tone of the papillary muscles. A most frequent and serious complication is atrial fibrillation (tachysystolic form) due to the toxic effect of the thyroid hormones on the myocardium. Circulatory insufficiency can also develop. Electrocardiograhic studies reveal a slightly increased amplitude of all waves (especially of the T wave), sinus tachycardia, extrasystole, and atrial fibrillation. X-rays examination reveals a slightly enlarged left ventricle of the heart.

Gastro-intestinal tract

The appetite increases. The increased motor function of the

intestine accounts for diarrhoea. Hepatic dysfunction can have various effects: from slight disorders (that can only be revealed by functional tests) to cirrhosis.

Nervous system

The clinical symptoms of disorders in the higher nervous activity are excitability, increased reactivity, general motor restlessness, fidgetiness, and fine tremor of the fingers of the stretched arms (*Marie's syndrome*).

Endocrine system

A pronounced clinical picture of the disease is attended by a marked hypofunction of the sex glands (*amenorrhoea*) and of the adrenal cortex (*hypoadrenocorticism*); diabetes mellitus can joint the process.

Study of the peripheral blood can reveal *hypochromic anaemia*, *leukopenia*, and *lymphocytosis*. Biochemical studies of blood reveal the tendency to *hypocholesterolaemia* and *hyperglycaemia*.

Basal metabolism increases by 50 and sometimes by 100 per cent. Tests with ^{13}I show accelerated and increased absorption of radioactive iodine by the thyroid gland, an increased content of protein-bound iodine, and decreased excretion of iodine in the urine. The body temperature is usually subfebrile.

Course

The course of the disease depends on the gravity of thyrotoxicosis and is divided into three degrees: I degree thyrotoxicosis is characterized by the absence of complications; wasting is not marked, tachycardia is moderate (to 100 beats per min), basal metabolism increases not more than by 30 per cent; the symptoms are pronounced in II degree of thyrotoxicosis (wasting is considerable, symptoms of nervous disorders are marked, *tachycardia* from 100 to 120 beats per min, basal metabolism increases by 30-60 per cent); III degree thyrotoxicosis: grave forms of the disease with pronounced symptoms (rapidly developing *cachexia*, marked psychic excitability and other nervous symptoms, pronounced tachycardia, over 120 beats per min, basal metabolism increased by more than 60 per cent). Forms of the disease complicated by atrial fibrillation, heart failure, affections of the liver, and psychoses are also referred to III degree thyrotoxicosis.

The main complications in *thyrotoxicosis* are affections of the internal organs, e.g. the heart or the liver, and also *psychoses*, *hypoadrnocorticism*, and *thyrotoxic crisis*.

Treatment

The patient should be given calm and rest; sleep should be normalized. The diet must be adequate, rich in proteins and vitamins. Antithyroid preparations should be given: iodine, thiouracyl, and imidazole derivatives. Transition from the second to the third degree is a positive indication for surgical intervention, irrespective of the length or gravity of the disease.

Hypothyroidism

Hypothyroidism is the pathological conditions associated with thyroid hypofunction. Hypothyroidism can be primary and secondary. Primary hypothyrodism is the primary pathology that arises in the thyroid gland, while secondary hypothyrodism depends on dysfunction of other organs that can affect the thyroid gland function. Grave forms of hypothyrodism are usually called myxoedema.

Aetiology and pathogenesis

Factor causing the onset of primary *hypothyroidism* are *hypoplasia* or *aplasia* of the thyroid gland, iodine deficiency in the body, subtotal thyroidectomy, overdosage of ^{13}I (which is given in hyperthyrodism) or preparations of thiouracil group, acute (in the past) or chronic thyroditis. Hyposecretion of thyroxine and triodthyrosine upsets normal metabolism and causes changes in tissues, organs, and systems of the body.

Pathological anatomy

Morphological changes in the thyroid gland are marked hypoplasia, aplasia, or atrophy. Hyperplasti changes occur in the thyroid gland in hypothyroidism caused by disordered synthesis of the hormones associated with the defective enzyme systems.

Clinical picture

The main complaints are apathy, lack of interest in the surroundings, impaired memory, decreased work capacity, somnolence, flaccidity, and chills.

The patients' appearance is quite specific: the eye slits are narrow, the face is puffy, the neck oedematous, the skin is pallid with a yellowish hue, sometimes with blush on the cheek bones. The skin is rough to the touch, thick, dry, cold, and scaling. The skin is thickened due to accumulation in if of mucopolysaccharides which give the impression of oedema. As distinct form oedema, pressure on the skin does not leave depressions. Hair on the head is rare; it falls off form the brows. Movements are slow and speech is monotonous.

Central and peripheral nervous systems

The mentioned complaints are associated with changes in the function of the central nervous system. Psychosis may develop in long-standing hypothyroidism. Disorders in the peripheral nervous system are manifested by strong severe radicular pain in the extremities, paraesthesia, cramps, and shaky gait.

Cardiovascular system

Bradycardia develops; the minute blood volume decreases and the blood flow rate is slow. The heart sounds are dulled. Fluid containing much protein and mucinous substances if often accumulated in the pericardium; it can be accumulated also in the pleural and abdominal cavity. Systolic pressure falls while diastoic pressure remains normal. ECG shows low voltage, especially in P and T waves. Heart failure develops in rare cases.

Gastro-intestinal tract Hyop-and achlorhydria often develops. The intestinal motor function is decreased, constipation and meteorism develop.

Metabolism

Protein synthesis is decreased. Blood cholesterol is usually increased. Moderate hypoglycaemia is observed. Electrolyte level remains unchanged in most cases. Blood calcium sometimes decreases, and ESR increases.

Reduction of basal metabolism to 50 per cent and also of the protein-bound iodine is of great diagnostic significance. Absorption of I^{131} in the thyroid gland is low.

Myxoedema coma may develop in grave cases.

Treatment

Thyroid preparations are mainly used to treat hypothyrodism and coma.

Diabetes Mellitus

Diabetes mellitus is characterized by metabolic disorders associated with absolute or relative deficiency of insulin production. *Diabetes mellitus* is a frequently occurring disease. People between the ages of 40 and 60 are mostly affected.

Aetiology and pathogenesis

Organic or functional affection of beta cells of the pancreas islets is the main factor in the pathogenesis of *diabetes mellitus*. This affection accounts for insufficient synthesis of insulin. Primary insufficiency of thee cells can arise after infection, psychic trauma, removal of the pancreas, its destruction by a tumour, sclerosis of the pancreatic vessels, in *pancreatitis*, regular overeating or insufficient intake of substances required for the normal function of the insular apparatus. Familial predisposition (genetically determined functional insufficiency of beta cells) is a background against which the diabetogenic effect of the named factors is realized.

Secondary insufficiency of beta cells can be due to endocrine dysfunction: pituitary, adrenal and thyroid hyperfunction. Somatotropic and thyrotropic hormones, corticotropin, gluccocorticoids and glucagon have diabaetogenic properties and are called contrainsulin hormones. The pathogenesis of diabetes mellitus also depends on the presence of excess insulin inhibitor, i.e. enzyme insulinase (which is produced in the liver and is activated in the anterior pituitary hyperfunction) and also insulin antagonists and antibodies to insulin contained in the blood of patients.

Hyperglycaemia is a symptom of disordered carbohydrate metabolism. Increased blood sugar content is associated with a slowed glucose supply to the muscles and fatty tissue and its slow phosphorylation. This interferes with glucose decomposition, synthesis of glycogen, and conversion of carbohydrates into fats. High blood sugar depends also on intensified glucose from glycogenic amino acids. Hyperglycaemia is usually attended by glycosuria, which in turn depends on an increased amount of glucose in the glomerular filtrate and its complete reabsorption in the tubules. Upset protein metabolism is manifested by the inhibited synthesis of protein. Clinically it is manifested by formation of trophic ulcers and slow heating of wounds.

Disorders of fat metabolism are delayed formation of higher fatty acids and neutral fats from carbohydrates, and ample supply of free fatty acids to the blood. Clinically this is manifested by wasting of the patient. Fat infiltration of the liver is the sign of upset fat metabolism. A severe disorder in fat metabolism is ketosis. This is an accumulation in the blood of acetone bodies and ketones (b-hydroxybutyric acid, acetoacetic acid, acetone) which are intermediate products of oxidation of higher fatty acids in the liver. Diabetic coma, a fatal complication of diabetes mellitus, can develop in this disorder of fat metabolism.

Polyria, loss of sodium and partially of potassium are symptoms of upset water-salt metabolisms diabetes mellitus. The pathogenesis of polyuria is associated with glycosuria which elevates osmotic pressure in the tubules to decrease reabsorption of water. Reabsorption of sodium in the kidneys is also decreased.

A long-standing and incompletely compensated diabetes mellitus results in vascular changes (retinopathy, nephropathy, or Kimmelstiel-Wilson syndrome) and atherosclerosis. Pronounced fluctuations in the blood sugar increase pituitary activity, cause spastic atonia of the vessels, which, in turn, affects the structure of their walls, accelerates the destruction of elastic fibres, and promotes sclerosis and calcinosis.

Insulin deficit inhibits phosphorylation of vitamin B6 which often causes neuropathic complications of diabetes mellitus.

Pathological anatomy

Diabetes mellitus is responsible for the decreased number of beta cells of the pancreatic islets, for their degranulation and hydropic degeneration. Hyaline and fat may be deposited in beta cells. This is not however a specific symptom of diabetes mellitus. At early stage of the disease, especially in young persons, morphological changes in these cells are absent.

Clinical picture

The symptoms of diabetes mellitus are excessive thirst (polydipsia), increased appetite, polyuria, hyperglycaemia, glycosuria, wasting, weakness, decreased work capacity, and skin itching, especially in the perineal region.

Inspection of the patient reveals rubeosis (reddening of the face, the cheeks, supracillary arches, and the chin due to dilated cutaneous vessels) and xanthosis (yellowish decolouration of the palms and soles associated with upset conversion of carotin into vitamin A in the liver and accumulation of carotin in the skin). The patient's skin is dry, rough, easily scaling, covered with traces of scratching (Due to skin itching). Furuncles, exematous and ulcerous lesions can also be found. At points of insulin injection, there are zones where fat is absent (insulin lipodystrophy).

Cardiovascular system

Atherosclerosis of various arteries with the corresponding clinical symptoms, angina pectoris, gangrene of the feet, etc. is not infrequent.

Respiratory organs

Diabetes mellitus often concurs with bronchitis, pneumonia, and pulmonary tuberculosis.

Gastro-intestinal tract. Mouth mucosa and the tongue are dry. Paradontosis and pyorrhea frequently occur. Appetite is very good and sometimes voracious (bulimia). Study of the gastric juice reveals the presence of hypo-or achlorhydria. Fat dystrophy of the liver and its cirhosis develop in some patients with long-standing decompensated diabetes mellitus.

Kidney diseases

Arteriolosclerosis of the kidneys and intracapillary glomerulosclerosis (Kimmelstiel-Wilson syndrome) may occur. They are manifested by hypertension, retinopathy, and albuminuria. Pyelonephritis is not infrequent.

Retinopathy

Retinopathy in diabetes mellitus is manifested by the presence of exudate in the retina, haemorrhages, and pigment abnormality in the yellow spot. Cataracts often occur.

Changes in the nervous system. *Polyneuritis is frequent. Headache, deranged sleep, and decreased work capacity are the symptoms of affection of the central nervous system.*

The main laboratory methods used to diagnose diabetes mellitus and assess its gravity are based on determination of sugar and ketone bodies in the urine, determination of sugar in the blood on a fasting stomach and during the day, and glucose tolerance tests.

When a patient suspected for diabetes mellitus is examined, his blood and urine are in the first instance tested for sugar. Sugar in the urine of a diabetes mellitus patient may be 5–8 per cent and more. Morning urine of patients with latent diabetes mellitus may be free from sugar, and daily urine should therefore be better studied. Urine taken after giving the patient a test meal or sugar can also be studied.

Blood of a healthy individual (With a fasting stomach) contains 4.4–6.6 mmol/l (80–120 mg/100 ml) of glucose. This concentration increases to 28–44 mmol/l (500-800 mg/100 ml) and more in diabetes mellitus patients. But in the mold forms of the disease the blood sugar may remain normal (especially so if the test is done on a fasting stomach). In such cases blood sugar should be determined 3 or 54 times a day with a normal diet given. If glycaemia appears to exceed normal in repeated glucose tolerance tests, the diagnosis of diabetes mellitus can be considered proved. After determining blood sugar on a fasting stomach, the patient is given to drink 50 g of glucose in 200 ml of water. Blood specimens are then taken at 30-minute intervals for 3 hours. The blood sugar in a healthy individual increases by about 50 percent (but not over 9.4 mmol/l or 170 mg/100 ml) during the first hour, while during the second hour the initial blood level is restored (or it may drop below normal). The rise in the blood sugar is higher in diabetes mellitus patients, the increase in sugar concentrations delayed, while the initial level is not restored even in three hours. There is a variant of the glucose tolerance test in which another portion of glucose is given to the patient in one hour following the first dose. The first glucose dose intensifies the secretion of insulin in healthy persons, and the second dose does not therefore increase the sugar concentration in the blood, while sugar curve of diabetes mellitus patients gives another ascent (Two-peak curve).

Glucose oxidase and Samogyi-Nelson tests are now used for determining blood sugar. The glucose oxidase method is used to determine true glucose of the blood and it is therefore most specific, but the normal glucose level is slightly underestimated compared with the Hagerdon and Jensen method (3.3–5.5 mmol/l, or 60–100 mg/100 ml). The sugar concentration in the blood depends also on the technique by which the blood specimen is taken: glucose level is higher in capillary than in the venous blood. Increased blood sugar does not always indicate diabetes mellitus since it may be the result of emotional excitation. *Glucosuria* is an indirect sing of hyperglycaemia. The presence of sugar in the urine in the absence of hyperglycaemia cannot be used as an evidence of diabetes mellitus either, since glucosuria can be due to decreased sugar permeability of the kidneys (renal threshold). In the presence of kidney pathology (nephrosclerosis), gluosuria may be absent even when the blood sugar is abnormally high.

Tests for urine sugar are qualitative and quantitative. Sugar can be determined in the urine by special indicator papers (glucotest) and tablets (for rapid determination of urine sugar). Determination of acetone and acetoacetic acid (acetone bodies) is obligatory. It should however be remembered that acetonuria can occur also in healthy individuals during fasting and in toxaemia of pregnancy.

Patients with clear signs of diabetes mellitus do not require glucose tolerance testing. Prednisolone or corticoglucose test should be carried out in persons predisposed to diabetes mellitus and with normal results of glucose tolerance test. The results of glucose tolerance test depend on various factors: fasting, pathological processes in the liver parenchyma, injuries, infections, acute disorders in cerebral circulation, and strong emotions.

Determination of the alkali reserve of the blood helps predict the approaching grave complication of diabetes mellitus, i.e. diabetic coma. The alkali reserve decreases sharply in moderate acidosis. It decreases not only in diabetes mellitus but also in acidosis of other aetiology, e.g. in fasting or in kidney disease.

Course

The onset of the disease may be acute or gradual. The first signs of diabetes mellitus may be persistent itching and furunculosis. By the course and severity of the symptoms, and also by the body response to the therapy given, the clinical picture of diabetes mellitus is differentiated into light, moderate, and grave. The degree of *hyperglycaemia*, *glycosuria*, the presence of ketone bodies in the urine, and the gravity of acidosis should also be taken into consideration. In addition to the mentioned forms of diabetes mellitus, the following three stages are distinguished in its course: prediabetes, masked diabetes, and true diabetes mellitus. Prediabetes can not be diagnosed by the existing methods. This can be defined as hereditary predisposition, obesity, and cases where newborns (Both dead and alive) weight over 4.5 kg. Masked diabetes mellitus can be detected by the glucose tolerance test. True diabetes mellitus is diagnosed by clinico-laboratory findings.

Diabetic coma is a grave and sometimes fatal complication of diabetes mellitus. It occurs if diabetes mellitus is treated improperly or if the disease is complicated by acute infections, injuries, or nervous stress. Toxic symptoms develop gradually in most cases and the onset of coma is preceded by its precursors (precomatose state). Excessive thirst develops along with polyuria, epigastric pain, dyspepsis, headache, and loss of appetite. The patient's breath smells of acetone (odour or rotten apples). *Precomatose* state is followed by the first phase of coma which is characterized (in addition to the mentioned symptoms which are gradually intensified) by a strong nervous excitement: insomnia, restlessness, clonic convulsions, and Kussmaul's respiration. The excitement is followed by a marked inhibition, the second phase of diabaetic coma: the patient develops dizziness, shows no interest in surroundings, and finally loses consciousness. When in a deep coma, the patient is motionless, the face may be pink or pallid, the skin dry, the muscle tone and tendon reflexes are decreased, pathological reflexes sometimes develop, the eyeball tone decreases, the eyeballs are soft to the touch, the pupils are narrow. Kussmaul's respiration is heard at a considerable distance. The pulse is low and fast; the arterial pressure falls. *Hypothermia*, *oliguria*, and sometimes *anuria* develop. Blood sugar markedly increases (from 22 to 55 μmol/l or from 400 to 1000 mg/

100 ml). The alkali reserve of blood decreases to 15-30 per cent (v/v), the number of ketone bodies increases along with increased content of non-protein (residual) nitrogen; the chloride content decreases. Leucocytosis in coma can be as high as 50×10^9 per 1 l of blood with a neutrophilic shift to the left. Ketone bodies and considerable amounts of sugar are found in the urine. But gradual development of diabetic coma and distinct stages of this process are not always observed, the terminal phase of diabetic coma may come suddenly, without precursors.

The pathogenesis of diabetic coma is associated with acidosis mainly on account of accumulation of ketone bodies and their toxic effect on the central nervous system.

Hypoglycaemis coma arises in patients treated with insulin for diabetes mellitus, if their diet lacks carbohydrates or as a result of insulin over-dosage. Hypoglycaemic coma develops rapidly, sometimes within a few minutes. Coma is preceded by a sudden feeling of hunger, weakness, sweating, tremor in the entire body, psychic and motor excitement. Comatose state is characterized by pallor and moist skin, increased muscular tone and tendon reflexes, and convulsions; the pupils are dilated, the eyeballs remain firm. The blood sugar is low; sugar and acetone are absent from the urine. The patient quickly responds to treatment: after an intravenous infusion of a hypertonic solution of glucose, the patient quickly regains consciousness.

Treatment

In the absence of malnutrition, ketosis, or concomitant diseases, and if there were no precomatose or comatose state, the patient may be given a diet therapy alone. Otherwise, and also if the antidiabetic preparations prove ineffective, or else if there are contraindications to their use, the patient should be given insulin.

The dose of insulin to treat diabetic coma depends on the gravity of the patient's condition and the length of the disease. In order to prevent the development of hypoglycaemic coma, a glucose solution in a hypertonic sodium chloride solution should be administered by drop infusion in 90–120 minutes following the administration of insulin.

Obesity

Obesity (adiposis) is excessive deposition of fat in subcutaneous and other tissues, which is associated with metabolic disorders.

Aetiology

Overeating is the main aetiological factor. Hypodynamia, hereditary and constitutional predisposition are also important. Pregnancy, lactation, and menopause are among other factors responsible for obesity in women.

Obesity can be regarded as an independent disease in cases with an excessive caloric intake (alimentary obesity). Obesity can also be a symptom of endocrine diseases (thyroid or pituitary dysfunction) or diseases of the central nervous system (infection, injury, tumour).

Pathogenesis

The main pathogenic mechanisms of obesity is dysfunction of the central nervous mechanisms, i.e the cerebral cortex and hypothalamic centres (The ventromedial and ventrolateral nuclei of the hypothalamus) that regulate fat and carbohydrate metabolism. The result of these disorders is upset equilibrium between the caloric intake and the amount of energy spent by a living body. The role of the endocrine factors and also changes in the local tissue metabolism with increased deposition and also changes in the local tissue metabolism with increased deposition of fat should also be considered in various types of obesity. Changes in insulin content in obese individuals are of special practical interest: at the early stage of obesity, these patients have hyperinsulinism, which is followed by hypoinsulinaemia on account of exhaustion of the insular apparatus in long-standing obesity. Hypoinsulinaemia impairs tolerance to carbohydrates which occur in most obese persons, and often cause diabetes mellitus. Obesity can thus be regarded as prediabetes.

Pathological anatomy

Fat deposition is more pronounced in subcutaneous tissue, omentum, around the kidneys, and in the mediastinum. In the epicardium, fat is mainly deposited at the apex of the heart and around its right chambers. Fat can grow into the depth of the heart to separate muscle fibres, which thus grow thinner. The liver is enlarged at the expense of fatty (adipose) infiltration, which also affects the pancreas. Fat loosens the pancreatic pparenchyma and causes atrophy of pancreatic islets.

Classification

Two types of obesity are distinguished, primary and secondary. Primary obesity includes alimentary obesity whose primary pathogenetic factors are unknown. Secondary obesity includes the following forms: (a) cerebral (hypothalamic obesity due to affection of the central nervous system, e.g. by a tumour, infection, or injury); and (b) endocrine obesity which is connected with dysfunction of the pituitary and thyroid glands, or the ovaries.

There is a type of obesity in which fat is deposited in tender tumour-like growths (lipomatosis).

Clinical picture

The clinical picture of obesity is quite varied, depending on the degree of the condition, length of a pathological process, and the presence of changes in other organs and systems.

The degree of obesity is determined by Broca's formula (weight individual = height - 100). The obesity is first degree if the patient's weight exceeds that calculated from the formula by 30 per cent; the second degree—from 30 to 50 per cent; the third degree —from 50 to 100 per cent; and the fourth degree of obesity is characterized by more than 100 per cent excess in weight. Individuals with the first and second degree of obesity do not complain of their disease. They attend the doctor only for aesthetic consideration. Patients with obesity of the third and fourth degree complain of dyspnoea (which develops first under considerable load and later during light exercise), fatigue, impaired memory, hyperhidrosis, flaccidity, constipation, and menstrual disorders. General inspection of the patient alone is enough to establish the diagnosis of obesity. The colour of the skin may be normal; the skin can also be pallid or hyperamic. White, red, or violet stripes (striae) can be seen on the skin of the abdomen and the thighs. Sometimes the skin sags together with the subcutaneous fat to resemble an apron. Because of hyperhidrosis, obese patients often have skin diseases, such as eczema, pyodermia, and furunculosis. The diaphragm is high and for this reason obese patients often develop bronchitis and pneumonia.

Long-standing and pronoucned obesity rovokes change in the cardiovascular system. Hypertension and atherosclerosis are frequent. These pathological changes and also mechanical factors (Accumulation of fat in the mediastinum, decreased respiratory excursions, high diaphragm) interfere with normal work of the heart and cause chronic circulatory insufficiency.

Patients with obesity have increased appetite. They develop tendency to constipation and meteorism. Cholelithiasis, cholecystitis, cholangitis, and acute pancreatitis occur in obese patients more frequently than in normosthenic individuals. Sex and pancreatic function decreases in obesity.

Treatment

Low-calorie diet is recommended to decrease the weight of obese patients to the normal level, after which the diet should be normalized.

The meals should be taken at regular intervals; the patient should lead an active life (remedial exercises). Treatment of obesity by starvation or by decreasing significantly the ration is not recommended on account of possible serious disorders.

Preparations decreasing appetite (anorexigenics) should be given to patients on a fasting diet who suffer severely form hunger.

Prophylaxis. Health education and sports are effective.

Vitamin Deficiency

Vitamins are low-molecular chemical compounds contained in foods. They are necessary to maintain biocatalysis of separate biochemical and physiological processes in a living body. Vitamins are synthesized in man and animals in insufficient quantity and therefore they should be taken with food.

If food lacks vitamins, its assimilation is disordered and symptoms of vitamin deficiency develop. Pathological conditions occurring in full absence of vitamins are called avitaminosis. Types of avitaminosis depend on the lack of particular vitamin (e.g. vitamin A, B_1, B_2, etc.) and are designated avitaminosos A, avitaminosis B_1, etc., respectively. If deficiency of vitamins is only partial, this condition is called hypovitaminosis, the insufficiency is usually functional.

Aetiology and pathogenesis

Depending on the cause of vitamin deficiency, two forms are distinguished: exogenous and endogenous. Exogenous or primary vitamin deficiency is caused by the low vitamin content of food. This is primary vitamin deficiency which develops in monotonous and irregular meals and imbalanced diet (With prevalence of carbohydrates, small amounts of animal proteins and fats, and in the absence of fresh vegetables and fruits).

Exogenous or secondary vitamin deficiencies caused by various factors. These may be disordered absorption of vitamins in the gastrointestinal tract (alimentary diseases), in helminthiasis, acute and chronic liver diseases, malignant new growths, leucoses, familial enzymopathies, endocrine dysfunction (thyroid or pituitary dysfunction), ingestion of foods containing antivitamin properties (thiaminase, avidin, etc.). Vitamin deficiency can develop in prolonged use of medicinal preparations with antivitamin properties (streptomycin, chloramphenicol, sulpha drugs, etc.). Vitamin deficiency can arise with adequate intake of vitamins but in the presence of increased demands for these substances, for example at high or low ambient temperature, during intense exercise, nervous or psychic overstrain, and in oxygen deficiency.

Main Clinical Symptoms of Diseases in Vitamin Deficiency

Lacking vitamin	*Clinical signs of hypovitaminosis*
Ascorbic acid (vitamin C)	Loose and bleeding gums, petechial lesions, dry skin
Nicotinic acid (vitamin PP)	Scarlet tongue, tender and cracked. Burning sensation in the tongue. Diarrhoea without mucus of blood. Skin scaling, hyper- pigmentation. Neurasthenic syndrome (irritability, insomnia). Muscular pain.
Pyridoxine (vitamin B_6)	Increased excitability, loss of appetite, nausea. Hypochromic anaemia. Seborrhoel dermatitis. Insomnia. Depression.
Retinol (vitamin A)	Palid and dry skin. Cornification of hair follicles. Acme. Tendency to pyodermia. Conjunctivitis. Photophobia. Night blindness. Brittle and reedy nails. Frequent respiratory affections.
Riboflavin (vitamin B_2)	Dry lips. Vertical fissures on the lips. Fissurs and crusts in the mouth angles (Angular stomatitis)). Seborrhoeal dermatitis of the face, ears, and the neck. Conjunctivities. Ready an dbrittle nails.
Thiamine (vitamin B_1)	Decreased appetite; nausea, constipation. Muscular dystonia. Distractedness, loos of self-confidence, non-motivated fears. Tenderm palpation of the calf lmuscles. Heart palpitation. Dyspnoea during light exercise. Rapid fatigue (mentl an dpsychic).
Tocopherol (Vitamin E)	Muscular weakness. Sex disorders.
Phylloquinone (vitamin K)	Nasal and gum bleeding. Intracutaneous and cutaneous haemorrhages. Gastro-intestinal haemorrhages.
Folic acid	See vitamin B12 (folic acid) deficiency anaemia
Cyanocobalamin (vitamin B_{12})	See vitamin B12 (folic acid) deficiency anaemia.

Demands in vitamins, especially for ascorbic acid, pyridoxine, folic acid, calcipherols, and tocopherols especially increase in pregnancy and nursing. Vitamin deficiency can develop also in prolonged use of a diet lacking vegetables, fruits, rye bread, etc. Avitaminosis develops during wars and other disasters, in long sea voyages, in polar regions where people feed on dry or canned foods, etc. Beri-beri (B_1 avitaminosis), pellagra (PP avitaminosis), scorbutus (scurvy, or C avitaminosis) occur now in developing countries usually suffers from hypovitaminosis, which arises due to improper storage of foods, seasonal variations in vitamin content of foods, or incorrect processing of foods.

Clinical picture

Pure hypovitaminosis occurs comparatively rarely. More frequent are deficiencies of various vitamins, with prevalence of symptoms characteristic for the lack of a particular vitamin. General vitamin deficiency develops in practically healthy individuals early in spring and it is manifested by rapid fatigue, decreased work capacity, poor appetite, and deranged sleep. Table gives clinical signs of separate hypovitaminoses in deficit of separate vitamins. Diagnosis of avitaminosis (in the presence of specific pathological symptoms) can be confirmed by testing the blood and urine for the presence of vitamins or their metabolities. Biochemical tests are also used to obtain indirect evidence of vitamin metabolic disorders. For example, tryptophan is given to a patient and excretion of xanthurenic acid in the urine is determined. This gives information of the pyridoxine supply of the body.

Determination of eye sensitivity to light helps establish deficiency of retinol and riboflavin. Immunobiological tests are also used: phagocytic reaction, complementary activity test, gamma-globulin concentration test, etc.

Radio-indicative method is also used to determine vitamin deficiency. It can be used to determine distribution, transport, conversion, and excretion of vitamins.

Treatment

Diet containing much vitamins and vitamin preparations (in doses several times exceeding normal daily demands) should be prescribed. Hypervitaminosis and toxic complications are however possible in prolonged therapy.

Prophylaxis. Health education of population is important. People should be taught to feed properly, and instructions should be given how to cook and store foods. General knowledge of vitaminology is useful. Special measures must be taken by the appropriate governmental institutions. These measures include increased production of foods rich in vitamins, and of preparations (vitamins and polyvitamins). Foods for sale should be stored and processed properly at public catering enterprises. Vitamin content of food should be increased by selection of agricultural crops and rational animal breeding.

Diseases of Bones, Muscles and Connective Tissue

MAJOR CLINICAL SYNDROME

Anaphylactic Shock

Anaphylactic shock is a symptom complex of acute grave general allergic reactions of immediate type, characterized mainly by the initial stimulation and subsequent inhibition of the function of the central nervous system, bronchospasm, and a marked arterial hypotension.

Aetiology

An anaphylactic shock may be caused by repeated intake of substances which sensitize the body when taken for the first time. Usually these are medicinal preparations such as penicilin, streptomycin, procaine, vitamin B1, some other antibiotics, sulpha drugs, vaccines, sera, extracts of pollen of some plants, etc. It is important to note that an anaphylactic shock may develop after administration of small doses of the preparation which was given earlier in larger doses, e.g. intracutaneous injection of only a few units of penicillin (in diagnostic test for allergy). An anaphylactic shock may develop from using a syringe which was sterilized together with syringes and needles that were used to inject penicillin to other patients. Inquiry of patients predisposed to anaphylactic shock can often reveal allergic reactions in the past history. An anaphylactic shock usually arises in parenteral administration of medicines but it can also develop when these substances contact the mucosa. In some cases, an anaphylactic shock may occur from insect bites.

Pathogenesis

The pathogenesis of anaphylactic shock consists in sensitization of the body during the first intake of the antigen (medicinal substance), vaccine, etc.) and production, which are partly fixed on various tsisue cells. ON repeated intake on the same substance, a rection occurs by which an antigen-antibody complex is produced. Biologically potent substances, such as histamine, bradykinin, serotonin, etc., are released immediately from the cells into the blood in large quantities.These substances produce various effects on the organs and systems of the body to cause spasms of smooth muscles and to increase vascular permeability, while combination of the antigen with the circulating antibodies activates the complement and causes formation of anaphylatoxin. In atopy (hereidtary allergy characterized by congenital presence of antibodies to certain allergens), an anaphylactic shock may develop during the first contact with this substance.

Clinical picture

In addition to the described geneal symptoms, anaphylactic shock may have some specifc features. Anaphylactic shock develops rapidly, in a few seconds or minutes (To 30 minutes), following the intake of the allergen. The first symptoms are usually vertigo, headache, fear, cold sweat, dyspnoea, anxiety, pressure in the chest, and paroxysms of cough. In some cases, skin itching develops simultaneously. Some patients develop allergic urticaria, allergic oedema, tachycardia, abdominal pian, vomiting, diarrhoea, and often convulsions. The further picture varies: rapidly developing oedema of the throat and asphyxia, progressive hypotony, oedema and haemorrhages into the internal organs (which are especially dangerous if they affect the brain). In grave cases, the patient soon loses consiousness; this is an unfavourable prognostic sign.

Despite the varied clinical picture of anaphylactic shock, its diagnosis is not difficult: the main sign is the rapid response of the patient to the administration of the medicine. The shock may occur immediately. A routine systemic examination of the patient is impossible and urgent measures should be taken to recover the patient from the shock. This done, the physician may proceed with verification of the diagnosis.

Prognosis

The prognosis is serious in all cases: the patient may die within the first minutes or hours of asphyxia. Cardiovascular insufficiency, or irreversible affectiosn of the vitally important organs. The latter may develop and become the cause of death at later terms (in several days). After the patient has been drawn from the critical state, he should be given a thorough medical observation and examination by laboratory and instrumental methods (as indicated). This enables the physician to diagnose the affection of this or that organ at the early stage of the process.

Treatment

It is necessary to stop the allergic effect, g.g. to apply a tourniquet to the extremity into which the medicine was injected or which was bitten by an insect. This should be followed by administration of adrenaline (as a vasoconstrictive agent) to arrest the allergen supply from the tissues into the blood. Antihistamine preparations (dimedrol, suprastin, etc.), glucocorticosteroids and their analogues (prednisolone etc.), havign pronounced anti-allergic and anti-inflammatory action, should also be given. Depending on the special character of each particular case, symptomatic treatment should be given: oxygen therapy, cardiac glycosides, angiotonics, etc.

Prophylaxis

A thoroughly collected allergic anamnesis is very important. The patient should be asked to what preparations he might have allergic response, or if he has atopy or hereditary predisposition to allergic reactions. If this information is available, the physician should exclude those preparations to which the patient has the allergic reaction. Any room, where patients are given injections of medicinal preparations, should be equipped with all necessary means to recover patients from possible anaphylactic shock.

Allergic Oedema

Allergic oedema (angioneurotic oedema, Quincke's oedema) is characterized by attacks of a transient circumscribed oedema of the skin, subcutaneous connective tissue, and mucos.

Aetiology and pathogenesis

This is an allergic reaction to various allergens. Vascular reactions, and in the first instance increased vascular permeability, are important factors causign allergic oedema.

Clinical symptoms

The angioneurotic oedema develops acutely, a few seconds or minutes following the intake of the allergen (usually without any precursors). As a rule, oedema affects the lip, the cheek, the eye, but it can also develop in any organ (oedema of the throat, stomach, etc.). The oedema persists from a few minutes to several hours. The size of the swollen area varies, but it rarely exceeds the size of the palm. The allergic oedema may recur, not infrequently on the same organ.

Treatament

Intravenous infusions of a 10 per cent calcium gluconate solutin, administration of antihistaminic preparations, glucocorticosteriods (prednisolone, etc.). Symptomatic therapy is also recommended (e.g. in oedema of the laryngeal mucosa).

Prophylaxis

Medicines or foods which are know to cause the allergic reactions should be excluded.

SPECIAL PATHOLOGY

Rhematoid Arthritis

Rhematoid arthritis (infectious allergic polyarthritis, infectious non-specific deforming polyarthritis) is a systemic disease of the joints, first and foremost of small joints. The incidence among women is higher than in men. Young and middle-aged individuals are usually affected. The incidence of the disease is rather high: according to various authors, from 0.8 to 5 per cent of the population are affected by the disease.

Aetiology and pahtogenesis

Aetiology and pahtogenesis are unknown. The onset of the disease is usually associated with the presence of chronic infectious foci (Streptococcal infection, unknown virus, possibly mycoplasm). The disease is regarded as an infectious and allergic and is referred to the group of larger collagen diseases. The rheumatoid factor (antibodies to the Fc fragment of immunoglobulin and mainly to the class M immunoglobulin) and antibodies to DNA, collagen, and to the formed blood elements ar regularly found in the blood of patients with rheumatoid arthritis. It is believed that the rheumatoid factor arises in response to the production of auto-antigens, the proteins of the affected synovial membrane of the joints. The reaction between the antigen and the antibody causes a progressive affection of the joints with firther development of pathological proteins (auto-allergens). Hereditary predisposition to the disease is also important.

Pathological anatomy

Osteosynovitis is characteristic of the initial period. Later the cartilages and periarticular tissues are involved in the process. Fibrous-sclerotic changes develop which finally result in complete or incomplete dislocation of the joints, development of ankylosis and marked deformation of the joints (henc another name: deforming polyarthritis). In addition to pronounced affections of the joints, connective tissue and vessels are also affected in various organs which makes ti possible to regard rheumatoid arthritis as a systemic disease.

Clinical picture

Polyarthralgia with mostly symmmetrical affection of minor joints of hand and feet, limitation of movement, which is specially pronounced after prolonged inactivity, and progressive deformation of the affected joints are symptoms characteristic of the disease. But all other joints can also be involved. In some cases, the disease is monoarthritis. The onset of the disease is usually subacute, but it may be acute or prolonged. Pain in the joints is especially severe in the morning; it decreases at rest. After a night sleep or prolonged inactivity, the movement are especially limited, and articulation is difficult (this symptom is explained by oedema of periarticular tissues). By the night, movements of the joints become easier. The patient may also complain of general fatigue, fever, indisposition, weakness, and loss of appetite.

Inspection of patients with pronounced arthritic changes reveals specific deformation of the joints, their subluxation and ankylosis. The most typical signs of the disease are deviation of the hand in the ulnar direction (seal's fin deformity), flexion contracture of proximal and hyperdistension of distal interphalangeal joints (button-hole deformity), flexion contracture of the metacarpohalangeal joint with hyperdistensionin the proximal and flexion in the distal joint (swam-neck deformity of the finger). Inspection is supplemented by palpation of the joints to determinetheir tenderness, the degree of limitation of active and passive movements.

The typical changes in tehhand of patients with rheumatoid arthritis are a peculiar "visiting card" of the disease. It is an important diagnostic sign. Changes in the other joints are difficult to differentiate from similar changes occurring in arthritis of other aetiology.

Palpation reveals a greate or lesser degree of muscular atrophy (Due to lack of exercise, i.e. atrophy due to inactivity and due to specific muscular affection.) Sometimes firm rheumatoid nodules, 0.5–1.5 cm in diamter, usually mobile and not adherent to the surrounding tissues, can be palpated in subcutaneous connective tissue around the elbow, over the ulnar bone, Achilles tendon, and in the occipital aponeurosis.

Lymphadenopathy, spleno-and hepatomegaly are observed in juvenile rheumatoid polyarthritis. IN 80 per cent of cases, rhemaoid arthritis occurs in the form of joint affections. The arthrovisceral form of the disease occurs less frequently. In this case, direct examination of the patients and laboratory-instrumental methods reveal changes specific for affections of various organs (subacute or chronic myocarditis, pleuritis, diffuse fibrosing alveolitis, glomerulonephritis, or amyloidosis of various organs, e.g. kidneys, liver, etc., which attend rheumatoid arthritis).

Laboratory studies reveal increased ESR (to 50–60 mm/h), not infrequently normochromic anaemia, and also positive non-specific biocchemical tests showing the activity of inflammatory processes such as dysproteinaemia (hypergammaglobulinaemia, increased alpha globulins and fibrinogen in the blood serum), and high seromucoid and C-reactive protein in the blood. Detection of the rheumatoid factor in the blood serum and synovial fluid is a more specific laboratory test for rheumatoid arthritis.

X-raying of the joints reveals their specific changes: osteoporosis of bone epiphyses, narrowing and erosion of the joint slit, formation of microcysts in the epiphyses, osteophytes growing by the articulation surfaces. Complete and incomplete dislocation, marked deformities of the joints, and also complete overgrowth of certain articular slits can be revealed at later stage of the disease.

Course

The disease is chronic and progressive. In most patients, it is characterized by periodic exacerbations (provoked by infections, overcooling, etc.) and remission. The patient can die of a concurrent amyloidosis, affection of the vaitally important organs (heart or renal failure), and also complications associated with prolonged (sometimes, uncontrolled administration of strong medicinal preparations. These complications may be perforation of steroid ulcer of the stomach, or hypertonic and diabetic complications in prolonged use of glycocorticosteroids and their analogues.

Treatment

Therapy includes (1) sanation of chronic infection foci (carious teeth, tonsillitis, sinusitis, etc.); (2) using non-steroid anti-inflammatory analgesics (Acetylsalicylic acid, butadione, bruphen, and the like); (3) using chloroquine derivatives (delagil, plaquenil); (4) using glucocorticosteriods and their analogues (prednisolone and the like) in grave cases; this should also be supplemented by local therapy (given into the joint cavity); (5) remedial exercises and physiotherapy (mainly thermotherapy); (6) in individual cses, surgical treatment (synovectomy).

Prophylaxis

Prophylactic measures arenot well developed. But it has been shown that timely diagnosis and sanation of chronic infection foci (sanation of the mouth, tonsils, ears, etc.) are very important.

Osteoarthrosis

Osteoarthrosis (Deforming arthrosis) is a chronic dystrophy of the joints and periartiular tissues which causes deformation of the joints. The disease usually attacks aged and middle-aged women.

Aetiology and pathogenesis

Aetiology and pathogenesis are not yet known. It is believed that osteroarthrosis is associated with metabolic disorders in the cartilage due to its premature ageing. General endocrine and metabolic disorders, chronic microtrauma of the joints, and hereditary predisposition are also important. Synovitis, the reactive inflammatin, is secondary to irritation of the synovial membrane by articular detritu (minute grains of necrotized cartilage). Compensatory irritation of the cartilage occurs also with formation of osteophytes; subchondral oseoschlerosis develops.

Pathological anatomy

Dystrophic changes in the cartilage impair its elasticity. The cartilage surface becomes dry, opaque and rough. In grave cases, the cartilage may be affected by necrosis and ulceration, deformation of the joint surface, formation of osteophytes, perichondral osteosclerosis, and finally secondary arthroses. Inflammation in the cartilage and periarticular tissues is usually mildly pronounced.

Clinical picture

The patient complains of pain during exercise (walking, stepping on the affected leg) in the spinal column, the large joints of the lower extremities, and in distal interphalangeal joints of the hands. The pain is abated at rest.

Deformation of the joints and swelling around them can only be revealed in grave forms of the disease. Palpation can reveal mild tenderness of tissues surrounding the affected joint. Movements of the joints are difficult only at the last stage of the disease. It concerns mostly the hip joint (coxarthrosis): patients develop waddling gait due to difficult articulation of the bones. As the disease progresses, walking becomes impossible because of dislocation or marked deformation of the hip joint. Laboratory studies do not reveal any significant changes, except a mildly increased ESR. X-rays reveal narrowing of the joint slit, the presence of osteophytes, subcartilaginous osteosclerosis in combiantion with cyst-like areas of diminished density in the epiphyses, deformities of the joints of various degrees, and in some cases dislocaton of the joint.

Course

The disease is chronic and gradually progressive. Dislocation of the damaged joint is the possible complication.

Treatment

Therapy includes prescription of (1) analgesic and anti-inflammatory prparations: (2) intra-articular injection of trasylol (inhibitor of the proteases that are involved in the degnerative changes in the cartilage) and arteparon (to inhibit splitting of cartilage mucopolysaccharides); (3) remedial exercises and physiotherapy (mainly thermotherapy).

Haemorrhagic Vasculitis

Haemorrhagic vasculitis is an immuno-allergic disease characterized by affection of the capillaires, minor blood vessels, with subsequent multiple haemorrhages. The disease was first described by Schenlein in 1832 and then by Henoch in 1868.

Aetiology

Aetiology of the disease is uncertain, but it has been long noticed that haemorrhagic vasculitis often attends certain infectious disease (influenza, tonsillitis, tuberculosis, etc.) or develops in the presence of hypersensitivity to some foods and medicines. Specific antibodies to enothelial cells of the vascular wall are found in the blood of patients with haemorrhagic vasculitis.

Pathological anatomy

There are multiple haemorrhages in the skin, the wall of the gastro-intestinal tract,and less frequently, in other organs. Histological studies of tissues in the zone of haemorrhage reveal affections of the capillaries and minor vessels, necrosis of their vascular wall, thrombosis, proliferation of the intima with narrowed lumen of the vessel, and focal perivascular infiltration.

Clincal picture

The disease is characterized by a sudden multiple haemorrhages in the skin, often at symmetrical points of the right and left part of the trunk and the extremities. The joints may be affected: this is attended by pain, limited movements, and swelling of the periarticular tissues. IN the abdominal formof vasculitis, haemorrhagic lesions appear on the gastro-intestinal and peritoneal mucosa; the patient complains of severe pain in the abdomen, which is attended by bloody vomitting and stools; the belly is tensed, the general condition grave, like in the acute abdomen syndrome. In certain cases, , haemorrhagic vasculitis can proceed with affections of the kidneys (subacute or chronic glomerulonephritis) and other organs.

In mild cases, there are no changes in the blood, but severe haemorrhage is attended by hypochromic anaemia. The thrombocyte and fibrinogen content of blood is normal. Blood coagulation, bleeding time and clot retraction time remain norma. The tourniquet and pinch tests, and also the Bittorf-Tushinsky symptom are positive in most cases.

Course

The disease may be acute and continue for several days or weeks, or it may be chronic, with periodic exacerbations. The patient may die of profuse bleeding or haemorrhage into vitally important organs; necrosis of the intestine and kidney affections can be other cause of death.

Treatment

Treatment should be symptomatic. Calcium chloride is given intravenously to cause a weak anti-allergic effect and to strengthen the vascular wall. Salicylates and butadion are given in cases with affected joints. Corticosteriods are effective. They decrease substantially the allergic response of the body. Dimedrol, diprazine and other antihistamine preparations are also efficacious.

INDEX

D

E

G

H

I

T

U

V

W